AF251985

WAGENINGEN AGRICULTURAL UNIVERSITY PAPERS
89-3 (1989)

A PHYSICAL THEORY OF FOCUS DEVELOPMENT IN PLANT DISEASE

M. W. Zawołek and J. C. Zadoks

231 p.

Cip-data Koninklijke Bibliotheek, Den Haag

ISBN 90-6754-159-1
NUGI 835

Printed in the Netherlands by Drukkerij Veenman B.V., Wageningen

ABSTRACT

Focus formation by plant pathogens dispersed by air-borne spores was studied by means of a `diffusion theory'. The theory was derived from a set of simple assumptions summarizing existing phytopathological knowledge. The theory is a subset of the Diekmann-Thieme theory. It is formulated as a system of two partial differential equations: (1) the diffusion equation and (2) the generalized Vanderplank equation. A `diffusion model' is built on the basis of the `diffusion theory' to simulate a particular situation, which can be programmed in FORTRAN and linked to PODESS (Partial and/or Ordinary Differential Equation Systems Solver). The theory was validated by comparison to other epidemiological models and to experimental data. Sensitivity analysis, of a type new in agriculture, examined the influence of various parameters on the model's output. The `diffusion model' permits dynamic simulation of focus expansion of air-borne plant disease in the horizontal plane and in a three-dimensional multilayer crop. Disease development in a non-uniform crop, generation of daughter foci, and focus development under the influence of wind can be simulated. The model shows convincingly that epidemic development from a single focus proceeds more efficiently if at least two dispersal mechanisms with different parameters concur; disease spread is most rapid when the partitioning of spores over the two mechanisms reaches an optimum value.

CONTENTS

PREFACE

Conversations between the authors, a physicist and a plant pathologist who met by chance, led to the question whether the toolbox of the physicist could contribute to better understanding of focus development in plant disease. The present volume answers that question.

A 'diffusion theory' was developed and a numerical 'diffusion model' derived. The diffusion model has the form of a dynamic simulation model, adaptable to a variety of situations. These situations should be limited, for the time being, to foci in the crop canopy caused by fungal plant pathogens dispersed by air-borne spores.

The development of the 'diffusion theory' does not stand alone. The mathematicians O. Diekmann and H.R. Thieme found an integro-differential equation for focus development. F. van den Bosch and others applied the Diekmann-Thieme theory to focus development in plant disease, making the general theory more specific. Along the ruler connecting the general and the specific, the 'diffusion theory' is placed near the specific end. The simulation model can handle parameter variation in time and space. It is specifically suited to situations with an inhomogeneous distribution of the host plants and with multiple dispersal.

Several institutions and individuals contributed to the present volume. The Institute for Plant Breeding and Acclimatization, Cereals Department, at Cracow, Poland, allowed the first author prolonged leaves of study in The Netherlands. The Food and Agriculture Organization of the United Nations, the Ministry of Agriculture and Fisheries of The Netherlands (by way of the International Agricultural Centre, Wageningen) and the Wageningen Agricultural University, The Netherlands, accorded fellowships to the first author. The Department of Phytopathology provided working space, library services, and computer facilities. A.J. Koster of the Computer Centre, Wageningen Agricultural University, was quite liberal in the accounting of computer time, not calculated in seconds but in days. The 'Agricultural University of Wageningen Papers' kindly accepted to publish this volume.

The version of this volume presented as a Ph.D. thesis was made possible by the Foundation 'Fonds Landbouw Export Bureau 1916/1918'.

The authors acknowledge with great pleasure the assistance of J.A.J. Metz, Professor of Mathematical Biology at the Leyden State University, The Netherlands, and his invaluable contribution to the mathematical aspects of the 'diffusion theory', especially his efforts to place the 'diffusion theory' in a wider mathematical context. F. van den Bosch and J.A.P. Heesterbeek showed their continued and stimulating interest.

LIST OF SYMBOLS

AREA	– area of a single lesion	$[L^2]$
b_0	– coefficient	$[1]$
b_i	– coefficient	$[1]$
b_{ij}	– coefficient	$[1]$
$\vec{c}_o$	– velocity of focus expansion	$[LT^{-1}, LT^{-1}]$
C	– macroscopic cross section	$[L^{-1}]$
C_a	– macroscopic cross section for absorption	$[L^{-1}]$
C_s	– macroscopic cross section for scattering	$[L^{-1}]$
D	– diffusion coefficient	$[L^2 T^{-1}]$
dA	– surface element	$[L^2]$
dr	– radius element	$[L]$
dx	– length element	$[L]$
dy	– length element	$[L]$
dz	– length element	$[L]$
$d\alpha$	– angle element	$[1]$
$d\theta$	– angle element	$[1]$
E	– effectiveness	$[1]$
f	– function	$[1]$
$F(\vec{r},\theta,t)$	– local density of spores at $\vec{r}$ and t moving in direction θ	$[NL^{-2}]$
G	– fraction of spores removed from epidemic	$[1]$
i	– infectious period	$[T]$
I	– probability of infection	$[1]$
$\Im$	– ratio of infectious to latency period	$[1]$
j	– length of $\vec{j}$	$[NL^{-2}T^{-1}]$
$\vec{j}(\vec{r},\vec{\Omega},t)$	– flux at $\vec{r}$ and t of spores flowing in the direction $\vec{\Omega}$	$[NL^{-2}T^{-1}, NL^{-2}T^{-1}, NL^{-2}T^{-1}]$
J	– length of $\vec{J}$ ‡	$[NL^{-1}T^{-1}]$

$\vec{J}(\vec{r},t)$ - spore density current at $\vec{r}$ and t [‡] $[NL^{-1}T^{-1}, NL^{-1}T^{-1}]$

J_x - x-component of $\vec{J}$ $[NL^{-1}T^{-1}]$

J_y - y-component of $\vec{J}$ $[NL^{-1}T^{-1}]$

J_{x+} - as J_x, but in the positive x-direction $[NL^{-1}T^{-1}]$

J_{x-} - as J_x, but in the negative x-direction $[NL^{-1}T^{-1}]$

$K(t-t')$ - number of spores produced at t per mother lesion created at $t-t'$, per unit of time $[NN^{-1}T^{-1}]$

L   - number $[1]$

L - lesion density [†] $[NL^{-2}]$

$\mathfrak{L}$ - total number of lesions present in the field $[1]$

L_{max} - maximum lesion density [†] $[NL^{-2}]$

LAI   - leaf area index $[1]$

N - number $[1]$

N_o - number $[1]$

$\mathfrak{N}$ - number $[1]$

p - latency period $[T]$

P - production term $[NL^{-2}T^{-1}]$

q - number of new effective spores per deposited effective spore $[NN^{-1}]$

r - length of $\vec{r}$ $[L]$

$\vec{r}$ - position vector $[L,L,L]$

$\vec{r}'$ - position vector $[L,L,L]$

R - number of spores produced by a sporulating lesion per unit of time $[NN^{-1}T^{-1}]$

$\mathfrak{R}$ - number of daughter lesions produced per sporulating mother lesion $[NN^{-1}]$

$s(\vec{r},\vec{\Omega},t)$ - volume density of spores at $\vec{r}$ and t flowing in the $\vec{\Omega}$

		direction	$[NL^{-3}]$
$S(\vec{r},t)$	–	spore density [†]	$[NL^{-2}]$
t	–	time	$[T]$
t'	–	time	$[T]$
U	–	linear size of a square field	$[L]$
$\mathfrak{u}$	–	ratio of field length to the width of the contact distribution	$[1]$
v	–	length of $\vec{v}$	$[LT^{-1}]$
$\vec{v}$	–	velocity vector	$[LT^{-1},LT^{-1},LT^{-1}]$
v_g	–	settling velocity	$[LT^{-1}]$
w	–	length of $\vec{w}$	$[LT^{-1}]$
$\vec{w}$	–	wind velocity	$[LT^{-1},LT^{-1},LT^{-1}]$
w_x	–	x-component of $\vec{w}$	$[LT^{-1}]$
w_y	–	y-component of $\vec{w}$	$[LT^{-1}]$
w_z	–	z-component of $\vec{w}$	$[LT^{-1}]$
x	–	spatial direction	$[L]$
x'	–	spatial direction	$[L]$
y	–	spatial direction	$[L]$
y'	–	spatial direction	$[L]$
z	–	spatial direction	$[L]$
z'	–	spatial direction	$[L]$
α	–	angle [*]	$[1]$
β	–	variable	$[T^{-1}]$
δ	–	rate of deposition	$[T^{-1}]$
λ_a	–	mean free path for absorption	$[L]$
λ_s	–	mean free path for scattering	$[L]$
λ_t	–	mean free path for transportation	$[L]$
λ_o	–	parameter	$[1]$
μ	–	coefficient	$[1]$
π	–	constant (3.14159)	$[1]$
θ	–	angle	$[1]$
ρ	–	mean value of the distance from the source of spores	$[L]$
σ^2	–	variance	$[L^2]$
τ	–	time	$[T]$

$\vec{\Omega}$ – a unit vector of a direction in space   [1]

Symbols of `local' use only are not mentioned in this list.

† at the three-dimensional case this density has dimension $[NL^{-3}]$

‡ in Chapter 6 $\vec{J}$ is the spore flux with dimension $[NL^{-2}T^{-1}, NL^{-2}T^{-1}, NL^{-2}T^{-1}]$ and J is the length of $\vec{J}$ with dimension $[NL^{-2}T^{-1}]$

* in Chapter 5 α is the distance to the `axial point'.

1 INTRODUCTION

A conspicuous and important phenomenon in plant disease epidemiology is the hot-spot or focus. It was defined by the British Mycological Society (Anonymous, 1953). Since 1963, the year when Van der Plank published his ideas on botanical epidemiology, mathematics occupies an important position in the description of focus development. But mathematics was used mainly as a tool to help with the formulation or generalization of experimental data (Gregory, 1968; Gregory, 1973; Aylor, 1978; Aylor, 1987). Computer simulation was used extensively (Kyosawa, 1976; Kampmeijer and Zadoks, 1977; Minogue and Fry 1983a, b). Mathematicians also made their contributions to epidemiology. Starting from Kermack and McKendrick's 1927 paper the work was continued by Kendall (1948, 1965), Mollison (1977), Diekmann (1978; 1979) and Thieme (1977; 1979). The Diekmann-Thieme theory in a certain sense is the most general theory of disease development in time and space. Starting from the minimum collection of assumptions, it already leads to the mathematical explanation of focus formation. The theoretical concepts of Diekmann and Thieme were put into a phytopathological context by Van den Bosch et al. (1988a, b, c).

The study presented here takes a different path, one which is usually followed by physicists. Its basis is a set of assumptions about the details of the underlying processes, which assume nothing but `common' knowledge. On this basis a theory is built. This is first of all a theory about interacting physical entities. In a mathematical sense, this theory is subsumed under the Diekmann-Thieme theory. The mathematical description is supported by a physical picture of the processes

involved. Moreover, and possibly even more important,
the fact that the theory is phrased in terms of
differential equations which are easy to solve
numerically makes it easy to study i.a. focus
development for short periods of time, inhomogeneous
space, and parameter dependence on time.

Having a proper description of focus development in
time and space, which is a characteristic element of
many epidemics, a great variety of situations met in
agricultural practice can be described. Thus the study
presented below can be summarized by the following
flowchart (Fig. 1.1).

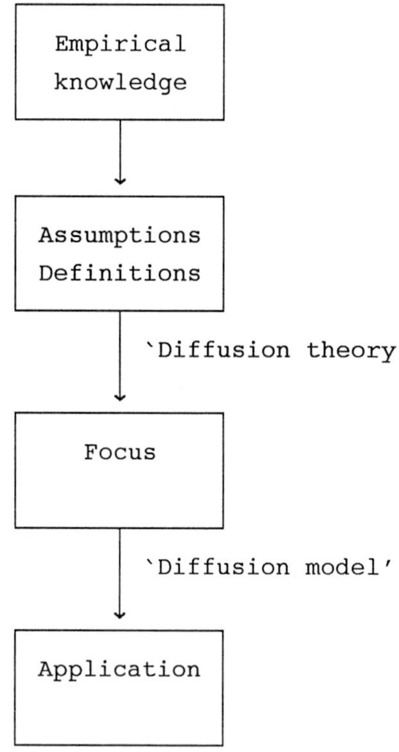

Fig. 1.1. Flowchart of the present study on focus
development.

2 OUTLINE OF THIS STUDY

The objective of this study is to present a theory, and practical applications of the theory, on focus development of plant disease in space and time. The book continues a tradition to formulate problems of botanical epidemiology in mathematical terms, as initiated by Van der Plank in his famous 1963 book and elaborated in e.g. EPIMUL by Kampmeijer & Zadoks (1977), in the Simulation Monographs. Analytical studies done in the late seventies by Diekmann (1978, 1979) and Thieme (1977, 1979), followed recently by the phytopathological interpretations of Van den Bosch et al. (1988a, b, c), shed new light on the application of mathematics to epidemiology. The aim of the present publication is to provide a theory, deeply rooted in physics and mathematics, which leads to models suitable for computer simulation of epidemics in time and space.

The theory borrows the conceptual approach from physics, modified where necessary. The language of this study is mathematics, but no great mathematical experience is required to understand the biological aspects of this study. As the theory deals with complex phenomena, analytical solutions are possible only in simple cases, usually too simple for application in agricultural practice. Therefore, numerical analysis and computer simulation are applied throughout. Some knowledge of FORTRAN is useful to understand and utilize the models presented, but the less experienced reader, who regards the programs as `black boxes', will be able to understand the results.

The theory is elaborated at several levels of complexity, gradually approaching the complexity of biological phenomena. Chapter 3 presents the `diffusion' model - based on the diffusion equation -

in two-dimensional space and time. Chapter 4 tries to validate the theory, comparing its results to the results of other models and to experimental data. Throughout chapters 3 and 4, the equations are solved either by analytical or by numerical methods.

Chapter 5 performs a sensitivity analysis in a way new to agriculture. Chapter 6 describes the 'telegrapher's' theory as a possible extension of the 'diffusion' theory. Comparison of the two theories indicates if it is worthwhile to use this extension. Another extension, incorporation of a wind effect on focus development, is described in Chapter 7.

The vertical distribution of spores and lesions, and the dependence of the distribution parameter values on height cannot be neglected. The models use parameters, which in reality are quite complex, so that they must be treated as empirical functions of time and environmental conditions. Stochasticity of some processes must be taken into consideration. Many diseases are spread by multiple mechanisms. Models, simulating these phenomena, are introduced in Chapter 8. Their realization using the principles of dynamic computer simulation is described in Chapter 9. Examples of the simulation models which show some of possible applications simulate inhomogeneity in crop distribution, multiple dispersal mechanisms, and three-dimensional crop distribution with developing leaf layers and variable wind.

Chapter 10 gives a general discussion with indications for future work.

This study is concluded by a brief description (Appendix A) of the computer package PODESS (Partial and/or Ordinary Differential Equations Systems Solver), written in FORTRAN, which is the base for the numerical treatment and the simulation programs. The flexibility of this package allows the user to run the programs discussed and to develop programs for his own needs.

3 DIFFUSION THEORY OF FOCAL DISEASE DEVELOPMENT
– THE APPROXIMATE THEORY FOR SPECIAL CASES

3.1 INTRODUCTION

The approach developed in this chapter describes disease development in time and space by means of equations which represent the governing processes. The solutions of the equations are functions, which are interpreted as densities of spores or lesions. This approach needs an assumption that these functions are continuous, so that the results are valid only for high lesion and spore densities. One cannot say exactly how many spores will be produced and liberated, where every one of them will be deposited, and whether the infection will be successful. But if one deals with high numbers of events, `acts of dispersal' as Vanderplank (1975) called them, the `average' behaviour of the spores can be determined by means of distribution functions associated with the process of dispersal. At the population level, spore dispersal can be treated as a deterministic process. Similarly, the other processes mentioned above can be treated as deterministic processes. The deterministic approach has proved especially fruitful in the description of spore dispersal proper (Gregory, 1973; Kiyosawa, 1976; Kampmeijer and Zadoks, 1977; Aylor, 1978; McCartney and Fitt, 1985; Van den Bosch et al., 1988a, b, c).

For low densities the stochastic nature of the process must be taken into consideration. If the density of spores is high and the chance of initialization of a lesion by a deposited spore is low, then spore production and dispersal are nearly deterministic, but the lesion production process is stochastic. A certain local spore density gives rise to infections according to a Poisson process with a rate

proportional to the product of the densities of
susceptibles and spores. Therefore, the function, which
in the deterministic approach is treated as the lesion
density, must now be interpreted as the mean number of
lesions initialized at a certain position (this number
itself is a variable with a Poisson distribution).

In this chapter a model of focus development in time
and in two spatial dimensions will be developed. The
theory presented is valid for short distance dispersal.
Plant height is neglected, and a crop is seen as a
horizontal plane. The starting point of the present
approach is the diffusion equation. Considering
comparable processes in different branches of science,
many authors derived the diffusion equation. They often
used different methods, but the result was always the
same (Okubo, 1980). As some information on its
derivation could be useful to phytopathologists (it is
good to know where and why one can use the diffusion
equation), a derivation will be given here. In addition
the situations considered in epidemiology are sometimes
different from those encountered in other branches of
science, so that a diffusion equation on itself is not
always satisfactory.

3.2 FOUNDATION OF THE THEORY

3.2.1 Definitions and assumptions

The first step into the world of equations, which
describe disease focus development in time and space,
is an exposition of definitions and assumptions, which
in strict statements capture the available knowledge
about the basic phytopathological phenomena. These are
the foundation of the theory of interacting entities:
sites and dispersal units. Because the assumptions are
nothing but a concise summary of empirical knowledge,
the theory which builds upon them should lead to a

proper description of focus formation (as confirmed in Chapter 4).

Definition D.1. A site is a limited area $[L^2]$ of host tissue, which can exist only in one of two states: non-infected (0) and infected (1).

In phytopathological literature, the terms 'occupied' and 'non-occupied' were used (Zadoks, 1971). The empirical counterpart of a site is a 'lesion' or the area (to be) occupied by it, therefore the two terms will be treated as equivalents.

Definition D.2. A dispersal unit is an entity, which can change the state of a site from non-infected to infected (0 → 1).

The empirical counterpart of the dispersal unit is a 'spore', therefore the two terms will be treated as equivalents.

Assumption A.1. The epidemic is described using two kinds of elementary units 'sites' and 'dispersal units'.

These elementary units demonstrate a set of properties. The properties allow to describe the behaviour of the units in a field situation.

Assumption A.2. The site properties are:
i.   *The transition between the classes of sites from non-infected to infected (0 → 1) is one-way only, and it is due to an outside influence.*
ii. *Once a site is infected, it is not a subject to an outside influence any more.*
iii. *A site from class 1 (infected site) can produce dispersal units.*

iv. Infected sites can belong to one of three subclasses: before production of dispersal units (subclass called 'latent', denoted 1A), during production of dispersal units (called 'infectious', denoted 1B), and after production of dispersal units (called 'removed', denoted 1C), allowing only a one way transition between them (1A → 1B → 1C).

v. Sites do not move.

vi. The density of sites is finite.

Ad. *A.2.iv*) The only subclass of infectious sites of which the contents cannot decrease in number is 1C (removed).

Ad. *A.2.vi*) The density of sites is limited by their size and by the leaf area index. For the deterministic theory it is necessary to assume that the number of sites in any finite area is infinite. In real life situations, the number of sites cannot be infinite, but a large number of sites allows the application of deterministic theory as a good approximation.

Corollary. C.1. The sum of the densities of sites in classes 0 and 1 is equal to the total density of sites. The sum of the densities of sites in subclasses 1A, 1B and 1C is equal to the density of sites in class 1.

Assumption A.3. The properties of dispersal units are:

i. Dispersal and deposition of a dispersal unit are governed by physical processes only.

ii. After its deposition on a non-infected site (0), a dispersal unit can, but does not necessarily, move the site to the infected class (1).

iii. A single dispersal unit can infect only one site.

vi. A deposited dispersal unit disappears.

v. There is no external source of dispersal units.

Ad. *A.3.i*) The dispersal unit can be moved (by air turbulences, wind, etc.) in a uniformly random direction, which changes continuously. During this movement, the dispersal units are deposited at a uniform rate.

Ad. *A.3.ii*) This change will be called `infection'.

Ad. *A.3.v*) The rate of change of the local density of dispersal units is equal to the rate of their local production minus their rate of emigration from a local area element plus their rate of immigration from neighbouring area elements minus their rate of deposition. The mathematical expression of this statement is:

$$\frac{\partial S}{\partial t} = \text{Production} + \text{Net migration effect} + \text{Deposition} \tag{3.1}$$

where S is the density of spores and t is time, $\partial S/\partial t$ is the rate of change of spore density S and `net migration effect' stands for `immigration' – `emigration'.

Definition D.3. An epidemic is a process, limited in time and space, which leads to an increase in the number of sites in state 1, substate C, at the cost of the number of sites in state 0.

3.2.2 Phytopathological context

The basis of the theory was formulated above. As the theory will be applied to focal spread of plant disease, a phytopathological context must be presented.

Spread of disease (sensu Vanderplank, 1975) takes place "when diseased plants occur where they did not occur before, either in the immediate past or at any time previously. The spread of disease implies the migration of pathogens." This is the phytopathological formulation of the spore properties *A.3.i*, and *A.3.ii* in assumption *A.3*. The words: host, disease, and pathogen are used as defined by the British Mycological Society (Anonymous, 1953).

The other spore properties are explained as follows. The spore properties *A.3.iii* and *A.3.iv* mean, that a deposited spore either germinates and infects (and then a single site becomes infected) or dies. There is no third possibility. Property *A.3.v* means, that an epidemic is treated as a closed system. The only source of spores is production by sporulating lesions, the only removal is by dispersal and deposition (in the language of physics `absorption').

The site properties are explained as follows. Properties *A.2.i* and *A.2.ii* mean, that an infected site cannot recover. Properties *A.2.iii* and *A.2.iv* are equivalent to the existence of non-zero duration of the latency and infectious periods (sensu Van der Plank, 1963). They lead to Vanderplank's or similar (as will be seen in Section 3.4.6) equations. Property *A.2.vi* is equivalent to the statement by Van der Plank (1963, p. 20) "As x increases, the proportion $(1-x)$ of susceptible tissue still healthy and available for infection decreases" (here x is the proportion of infected tissue density, i.e. the ratio of the density of infected sites to the density of all sites (vacant + infected). The word `density' was added to Vanderplank's explanation, because here we deal with space dependent models).

3.2.3 Terminology

The present theory requires its own terminology, which is introduced here, with symbols of variables and parameters (in italics) and their dimensions (in square brackets):

1. x = the first space coordinate   [L]

2. y = the second space coordinate [L]

3. t = time [T]

4. $\vec{r}$ = (x,y) - a position vector in 2-dimensional space (see below) [L,L]

5. $\vec{v}$ = velocity of spores (see below) $[LT^{-1},LT^{-1}]$

6. $\vec{c}_0$ = velocity of focus expansion (see below) $[LT^{-1},LT^{-1}]$

7. S = $S(\vec{r},t)$ - area density of spores at $\vec{r}$ and t $[N_s L^{-2}]$

8. L = $L(\vec{r},t)$ - area density of lesions at $\vec{r}$ and t $[N_L L^{-2}]$

9. F = $F(\vec{r},\theta,t)$ - local density of spores at $\vec{r}$ and t moving in direction θ $[N_s L^{-2}]$

10. $\vec{J}$ = $(J_x(\vec{r},t),J_y(\vec{r},t))$ - spore density current at $\vec{r}$ and t; it is the net number of spores per unit of time flowing through a unit of length perpendicular to the x- and y-direction) $[N_s L^{-1} T^{-1}, N_s L^{-1} T^{-1}]$

11. C = the macroscopic cross section for a given process (see below) $[L^{-1}]$

12. λ_s = the mean free path for scattering; $\lambda_s = 1/C_s$ where C_s is the macroscopic cross section for scattering [L]

13. λ_a = the mean free path for absorption; $\lambda_a = 1/C_a$ where C_a is the macroscopic cross section for absorption [L]

14. D = diffusion coefficient $[L^2 T^{-1}]$

Ad. 4. The value (length) of the vector $\vec{r}$ will be denoted r [L]; $r = |\vec{r}|$.

Ad. 5. The value (length) of the vector $\vec{v}$

will be denoted v [L]; $v = |\vec{v}|$.

Ad. 6. The value (length) of the vector $\vec{c}_o$
will be denoted c_o [LT^{-1}]; $c_o = |\vec{c}_o|$.

3.3 DIFFUSION EQUATION

3.3.1 Introduction

The following derivation of the diffusion equation
is based on the method used in the theory of nuclear
chain reactors (Glasstone and Edlund, 1956). Roughly
speaking, the nuclear chain reaction is the process of
production of new neutrons by nuclei which absorbed old
neutrons. During their movement, neutrons are scattered
and absorbed by nuclei at standstill. Absorption of a
neutron by a nucleus causes division of this nucleus
with production of energy and 2 or 3 new neutrons. The
process of plant disease focus development is rather
similar. Here and there, we deal with moving particles
which randomly change direction of their movement, are
removed from the population, and which can multiply
their numbers by reaction. Of course, the differences
in mechanisms involved in the two processes are
considerable. During their flight, spores follow air
turbulences rather than move along straight lines and
then suddenly change the direction of their movement by
scattering. Removal of a spore from the population is
not due to absorption but to deposition, which takes
place because local eddies around leaves are
sufficiently small that frictional drag is less than
inertia, so that spores move towards the surface
instead of being sidetracked with an air flow.
Generally, three processes are involved in the movement
of spores: gravity, air turbulence, and inertia.
Gravity can usually be neglected. When air speed
becomes sufficiently high, spores leave their host
plant and follow air turbulences. In general, air drag

dominates inertia except when the air flow changes direction very fast. Fast changes happen only in the transition layer from the boundary layers of the leaves.

Forgetting for some time the differences mentioned and keeping in mind a simplified picture of spore movement will lead to notions and equations which adequately describe plant disease development.

In the beginning, only spore dispersal will be considered. The deposition and production processes will be discussed later.

3.3.2 Continuity equation

The continuity equation is an elementary concept in physics, used here as a starting point. For simplicity, the continuity equation will be derived for spore movement in the x-direction only. After this simplified derivation, results will be generalized to an arbitrary-direction motion.

Imagine a small area element $dA = dx\ dy$, where dx and dy are small line elements in the x and y directions (Fig. 3.1). The element dA embraces a point with coordinates $\vec{r} = (x,y)$. A stream of spores flowing in the positive x-direction can be described by the x-component of the spore density current $J_x = J_x(x,y,t)$,

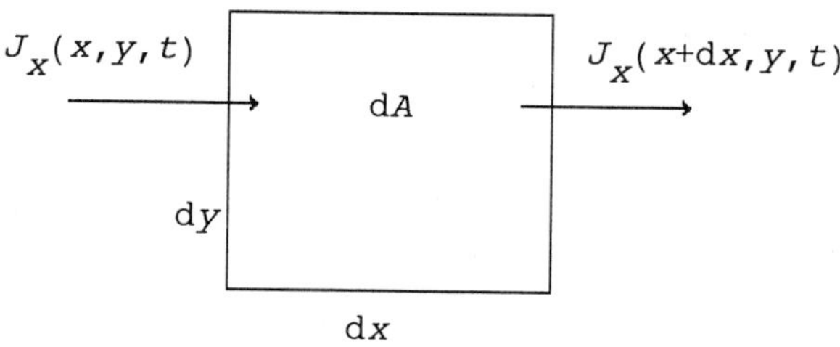

Fig. 3.1. Spore density current J flows through the area element dA.

the number of spores flowing in the x-direction and crossing a unit length during a unit of time $[NL^{-1}T^{-1}]$. The number of spores which flow into dA, during the time dt (dt being a short period of time), from the left side with position x is equal to the spore density current's x-component at x, $J_x(x,y,t)$, multiplied by the length of the left side of dA, dy and the time-period dt. Therefore, the number of spores flowing into to dA is:

$$N_{in} = J_x(x,y,t)\ \mathrm{d}y\ \mathrm{d}t \qquad (3.2)$$

The number of spores, which flow out of dA during dt through the right side of dA (this side is at the position x+dx) is equal to the spore density current's x-component at x+dx, $J_x(x+\mathrm{d}x,y,t)$, multiplied by the length of the right side of dA, dy and by the time-period dt. Therefore the number of spores leaving dA is:

$$N_{out} = J_x(x+\mathrm{d}x,y,t)\ \mathrm{d}y\ \mathrm{d}t \qquad (3.3)$$

At time t, the number of spores inside dA was $S(\vec{r},t)$ dA, where $S(\vec{r},t)$ is the area density of spores (treated as a constant in dA, because dA is very small) at time t. At time t+dt, the number of spores inside dA was $S(\vec{r},t+\mathrm{d}t)$ dA. Therefore, during dt, the change of the number of spores in dA equals $S(\vec{r},t+\mathrm{d}t)$ dA $-$ $S(\vec{r},t)$ dA on one hand and $N_{in} - N_{out}$ on the other hand. Then using (3.2) and (3.3) the following equation is formed:

$$(S(\vec{r},t+\mathrm{d}t) - S(\vec{r},t))\ \mathrm{d}x\ \mathrm{d}y =$$
$$(J_x(x,y,t) - J_x(x+\mathrm{d}x,y,t))\ \mathrm{d}y\ \mathrm{d}t \qquad (3.4)$$

where dA was replaced by dx dy. Dividing both sides of (3.4) by dx dy dt and passing to the limits dt = 0 and dx = 0, equation (3.4) becomes:

$$\left[\frac{\partial S(\vec{r},t)}{\partial t}\right]_m = -\frac{\partial J_x(\vec{r},t)}{\partial x} \qquad (3.5)$$

If spores move in an arbitrary direction, the analogs to equation (3.5) can be derived for each of the x- and y-directions. Therefore, the rate of change of the spore density due to movement in an arbitrary direction is:

$$\left[\frac{\partial S(\vec{r},t)}{\partial t}\right]_m = -\frac{\partial J_x(\vec{r},t)}{\partial x} - \frac{\partial J_y(\vec{r},t)}{\partial y}$$

$$= -\nabla \cdot \vec{J}(\vec{r},t) \qquad (3.6)$$

where $\nabla\cdot$ is an abreviated notation for the two terms appearing in the middle. ∇ is the differential operator (called `nabla'):

$$\nabla = \frac{\partial}{\partial x}\,\vec{i} + \frac{\partial}{\partial y}\,\vec{j} \qquad (3.7)$$

with $\vec{i}$ and $\vec{j}$ as the unit vectors in the x- and y-direction, respectively. In physics, equation (3.6) is called the `continuity equation'. In common words this equation means:

If there is no production or absorption of spores, the local rate of change in the density of spores will be due to the local difference between inflow and outflow.

3.3.3. From Fick's law to the diffusion term

Assume that spore movement is independent over infinitesimally small time and space scales. In that case net movement in the x-direction over a length element dy perpendicular to the x-direction, $J_{x'}$ equals:

$$J_x = \frac{D}{dx}\, S(x,y,t) \; - \; \frac{D}{dx}\, S(x+dx,y,t) \tag{3.8}$$

where D is a proportionality coefficient. The first
term refers to the movement from left to right and the
second term to the movement in opposite direction (Fig.
3.2). If spores move purely at random, both terms of
(3.8) are proportional to S and the division by dx
results from the assumption that if a spore is nearer
to x, the chance that it will cross the line segment dy
in x, in the nearest infinitesimal time interval,
increases. This increases as $1/dx$, because the result
should be finite and non-zero.

Passing to the limit $dx = 0$ and using the definition
of the first derivative, (3.8) becomes:

$$J_x = - D \, \frac{\partial S(x,y,t)}{\partial x} \tag{3.9}$$

Analogous to (3.9), the equation for the y-component,
$J_{y'}$ can be established. Finally

$$\vec{J} = (J_x,\; J_y) = -D \left[\frac{\partial}{\partial x}\, \vec{i} + \frac{\partial}{\partial y}\, \vec{j} \right] S$$

$$= -D \, \vec{\nabla}\, S \tag{3.10}$$

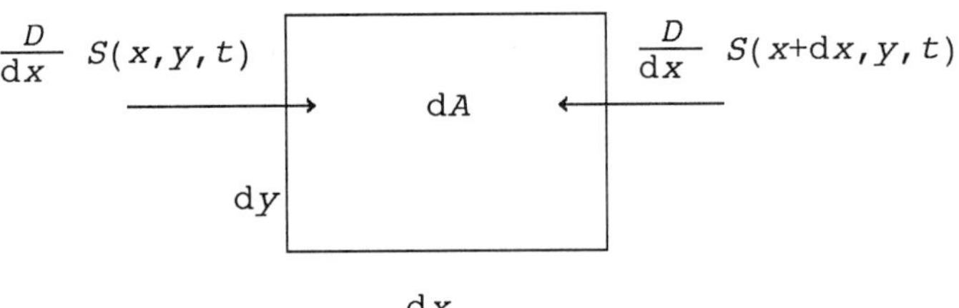

Fig. 3.2. Spore stream flows through the line element
dy at x and at $x+dx$ points.

where ∇S is the gradient of S. Equation (3.10) is known as Fick's law (Okubo, 1980; Hallam and Levin, 1986). In common words it means:

The net spore density current is proportional to the gradient of the spore density i.e. the net stream of spores flowing between two points is proportional to the difference between spore densities at these two points.

Substitution of (3.10) into the continuity equation (3.6) results in:

$$\left[\frac{\partial S}{\partial t}\right]_d = D\left[\frac{\partial^2 S}{\partial x^2} + \frac{\partial^2 S}{\partial y^2}\right] = D\,\nabla^2 S \qquad (3.11)$$

where ∇^2 is the Laplace operator. The subscript d indicates that the rate of change of spore density is due to diffusion (dispersal). Equation (3.11) is known as the 'diffusion equation' and D as the 'diffusion coefficient' (Okubo, 1980; Hallam and Levin, 1986). This form of the diffusion equation is applicable to the situations without production and deposition of spores.

3.3.4 A note on the scattering process

To give some notion what 'random and infinitesimally small' really means and therefore to establish the range of applicability of the diffusion equation (3.11), equation (3.9) will be derived once more. This derivation is based on a particular process, simple and exemplary, which is not random on an infinitesimally small scale. It is not intended as a proper discription of real turbulent diffusion but rather as a didactical example to sharpen the reader's intuition.

Imagine a small segment dl on the y-axis around the origin and a small surface element dA around a point with polar coordinates (r,α), where r is the radius (the distance from the origin of the coordinate frame to the point) and α is the angle between $\vec{r}$ and the x-axis (Fig. 3.3). Let $F(x,y,\theta,t)$ denote the local density of spores at (x,y) moving in the θ direction. The x-component of the spore density current can be expressed as:

$$J_x = J_{x+} - J_{x-} \qquad (3.12)$$

where the spore density currents are defined: J_{x+} is the number of spores crossing a unit segment (perpendicular to the direction of flow) per unit time in the positive x-direction, J_{x-} is the number of spores crossing a unit length (perpendicular to the direction of flow) per unit time in the negative x-direction.

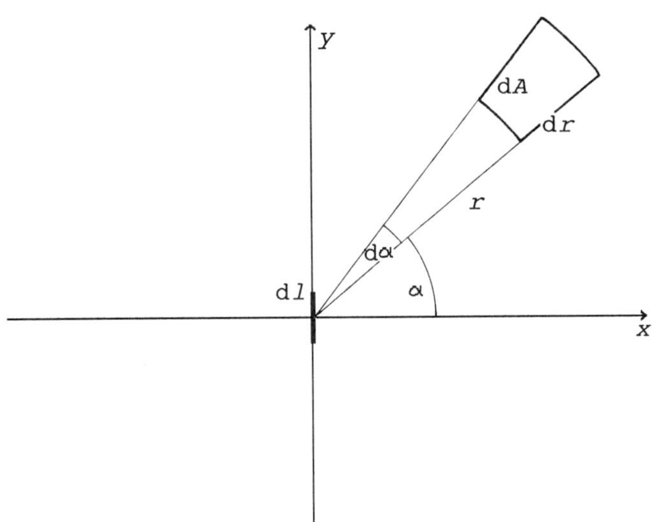

Fig. 3.3. Spores are scattered in dA in all directions. Some spores can pass through dl. How many (on average)? Explanation in text.

The density current of spores flowing from the right hand half-plane is:

$$J_{x-} = v \int_{-\pi/2}^{\pi/2} F(0,0,-\alpha,t) \, \cos(-\alpha) \, d\alpha \qquad (3.13)$$

where v is the spore velocity. To calculate $F(0,0,-\alpha,t) \cos(-\alpha)$ we look at the last scattering event experienced by a spore before it passed through dl. Assume that this scattering happened r away from the origin in dA around a point whose Cartesian coordinates are (x,y). They can be expressed in polar coordinates (r,α):

$$x = r \, \cos\alpha,$$
$$y = r \, \sin\alpha.$$

Therefore, a spore experienced its last scattering event at position $(r \cos\alpha, \; r \sin\alpha)$ at time $t-r/v$ (where v is the spore velocity). At this time there happened in an infinitesimal area dA

$$C_S \int_0^{2\pi} F(r \cos\alpha, r \sin\alpha, \theta, t-r/v) \, d\theta \, dA \qquad (3.14)$$

scattering events. C_S is a proportionality constant called the macroscopic cross section for scattering; it measures the 'intensity' of scattering (the concept of the macroscopic cross section is better explained in Appendix B; formula (3.14) is the two-dimensional counterpart of (B.1)). Assuming that the scattering is isotropic, the outflow from dA in every direction is equally probable. The probability that a spore will be scattered in such a direction that it will pass through the segment dl is:

$$\mathfrak{P} \; = \; \frac{\cos\alpha \cdot dl}{2 \cdot \pi \cdot r} \tag{3.15}$$

because $\cos\alpha\; dl$ is the projection of dl on a normal to the direction of flow, and $2\,\pi\,r$ is the length of the circle with its centre in dA and with radius r. Because of continuous scattering of spores on their way to dl, only the fraction $\exp(-C_S\, r)$ of them will arrive at dl without having undergone another scattering event (see equation (B.3) in Appendix B). The product of (3.14), (3.15) and $\exp(-C_S\, r)$ integrated over all radii from 0 to infinity, gives the flux of spores moving in the $-\alpha$ direction flowing through dl at t

$$F(0,0,-\alpha,t)\;\cos(-\alpha) \; = \; \frac{C_S}{2\pi} \int\limits_{0}^{\infty} \int\limits_{0}^{2\pi} F(r\cos\alpha, r\sin\alpha, \theta, t-r/v)$$

$$d\theta \; \cos\alpha \; \exp(-C_S\, r) \; dr \tag{3.16}$$

as in polar coordinates $dA = r\; dr\; d\alpha$. Substitution of (3.16) into (3.13) leads to

$$J_{x-} \; = \; \frac{v\cdot C_S}{2\pi} \int\limits_{0}^{\infty} \int\limits_{-\pi/2}^{\pi/2} \int\limits_{0}^{2\pi} F(r\cos\alpha, r\sin\alpha, \theta, t-r/v)$$

$$d\theta \; \cos\alpha \; d\alpha \; \exp(-C_S\, r) \; dr \tag{3.17}$$

Assume that both v and C_S are very large; in the limit case $v \rightarrow \infty$, $C_S \rightarrow \infty$, but in such a way that $v/C_S \rightarrow 2D$, where D is the diffusion coefficient. Then, (1) $t-r/v$ may be replaced by t, and (2) since the term $\exp(-C_S\, r) \rightarrow 0$ for other than infinitesimally small values of r, only the values of F near to the origin must be considered. Therefore, the spore density, F, can be expanded into a Taylor series. Restricting this expansion only to the first order terms, the expansion

26

can be written:

$$F(x,y,\theta,t) = F_0 + x \left[\frac{\partial F}{\partial x} \right]_0 + y \left[\frac{\partial F}{\partial y} \right]_0 + \ldots \qquad (3.18)$$

where the subscript 0 refers to the evaluation at the origin i.e. at the element dl. Introduction of (3.18) into (3.17) leads to:

$$
\begin{aligned}
J_{x-} = \frac{v \cdot C_s}{2\pi} \int_0^{2\pi} \Bigg[&F_0 \int_0^\infty \int_{-\pi/2}^{\pi/2} \exp(-C_s\, r)\, \cos\alpha\; d\alpha\; dr\; + \\
&\left[\frac{\partial F}{\partial x} \right]_0 \int_0^\infty \int_{-\pi/2}^{\pi/2} r\, \exp(-C_s\, r)\, \cos^2\alpha\; d\alpha\; dr\; + \\
&\left[\frac{\partial F}{\partial y} \right]_0 \int_0^\infty \int_{-\pi/2}^{\pi/2} r\, \exp(-C_s\, r)\, \sin\alpha\, \cos\alpha\; d\alpha\; dr \Bigg]\; d\theta
\end{aligned}
$$

$$\qquad (3.19)$$

After evaluation of these integrals, the spore density current in the negative x-direction can be expressed as:

$$J_{x-} = \int_0^{2\pi} \left\{ \frac{v \cdot F_0}{\pi} + \frac{v}{4 \cdot C_s} \left[\frac{\partial F}{\partial x} \right]_0 \right\} d\theta \qquad (3.20)$$

The equation for the spore density current, J_{x+}, can be derived after similar calculations (but integration over the angle α is now from $\pi/2$ to $3\pi/2$, and the sign is changed, because the current flows in the positive x-direction):

$$J_{x+} = \int_0^{2\pi} \left(\frac{v \cdot F_0}{\pi} - \frac{v}{4 \cdot C_s} \left[\frac{\partial F}{\partial x} \right]_0 \right) d\theta \qquad (3.21)$$

Substitution of (3.20) and (3.21) in (3.12) leads to

$$J_x = -D \int_0^{2\pi} \left[\frac{\partial F}{\partial x} \right]_0 d\theta \qquad (3.22)$$

where $D = v/2C_s = v \lambda_s/2$.

The number of scattering events experienced by a spore during a small time interval goes to infinity under the limiting operation described above ($v \to \infty$, $C_s \to \infty$). After a scattering event a spore necessarily has a random orientation of its movement, in a limit

$$F = \frac{S}{2\pi} \qquad (3.23)$$

Substitution of (3.23) into (3.22) leads to:

$$J_x = -D \frac{\partial S(x,y,t)}{\partial x}$$

This equation is identical to (3.9).

3.3.5 Diffusion formulation of the balance of spores

Apart from the diffusion term, the balance equation (3.1) contains absorption and production terms. These three terms together constitute the diffusion equation for dispersion, production and deposition of spores.

Assuming that airborne spores are deposited with probability δ per unit of time, the rate of change in the number of spores per unit area due to absorption is expressed by the following equation:

$$\left[\frac{\partial S}{\partial t} \right]_a = -\delta\, S \tag{3.24}$$

where subscript a indicates absorption. The proportionality constant, δ, is called the rate of deposition (absorption) $[T^{-1}]$.

The population of spores flowing through an absorbing medium with velocity v decreases proportionally to the distance passed and to the density of absorbing places. Therefore, the rate of change of spore density per unit of time should be proportional to v and to the macroscopic cross section for absorption, C_a (Appendix B). Equation (3.24) can be written in our particular scattering model as:

$$\left[\frac{\partial S}{\partial t} \right]_a = -v\, C_a\, S = -\frac{v}{\lambda_a}\, S \tag{3.25}$$

where λ_a – the mean free path for absorption can be defined as the distance at which $1/e$ spores is not yet absorbed.

Denoting the rate of spore production per unit area and per unit of time as P (the production term, $[NL^{-2}T^{-1}]$) and substituting this term together with the equations (3.11) and (3.25) into equation (3.1), the following `balance' equation for the rate of change of the number of spores per unit area is written:

$$\frac{\partial S}{\partial t} = D \left[\frac{\partial^2 S}{\partial x^2} + \frac{\partial^2 S}{\partial y^2} \right] - \delta\, S + P \tag{3.26}$$

Equation (3.26) is also called the `diffusion equation'. As written, it applies to the combination of dispersion, production and deposition of spores, but only when scattering is strong relative to production and deposition.

3.3.6 Discussion

The derivation of the diffusion equation presented above emphasized aspects relevant to phytopathologists. In addition to the assumptions stated in Section 3.2.1, other assumptions (purely random movement at an infinitesimally small scale, isotropy of space, and fast scattering) were made on the way. These assumptions permitted the derivation of the diffusion equation, but their consequences somewhat limit the range of applicability of this equation (they will be discussed in Section 3.5).

The relation between the physics of spore dispersal and both the diffusion and the scattering model presented above may pose a problem. The concept of mean free path for scattering cannot be applied directly to spore movement, because spores follow air turbulences rather than move along straight lines and change their direction of movement by scattering. Yet, spore movement is not purely random; there is some persistence in their movement due to inertia. Therefore, the mean free path for scattering is defined here as "the distance passed when $1/e$ spores move in a direction effectively independent from the old direction of spore movement". This parameter is identical to the 'mixing length' (Goudriaan, 1977), which characterizes air eddies. To strike a balance between theory and practice the following rule should be applied in the application of the diffusion equation:

Use the diffusion equation to describe focus development, if the mean free path for scattering is much shorter than the mean distance travelled by a spore during the period of interest, and if the mean free path for scattering is much shorter than the mean free path for absorption. Thus, the diameter of the 'solution' region must be considerably smaller

*than the distance travelled by a spore during the
time-period of interest.*

The limitations to the application of the
'diffusion' theory will be carefully considered in
Section 3.5. Keeping in mind the above limitations, we
can begin to solve the problem of disease focus
expansion by means of the 'diffusion theory'.

3.4 SOLUTIONS

3.4.1 Introduction

The solutions of the diffusion equation (3.26)
depend on the form of the production term P. In the
following pages solutions for different forms of this
term will be discussed. These solutions will be given
for a gradually growing complexity of the production
term, following the stepwise refinement of the models
by Van der Plank (1963). Therefore, not all the
solutions presented below refer to a phytopathological
reality. They are presented mainly as introductions to
the final, most complex, case. The latter is the only
phytopathologically relevant case. The reader may view
some of these solutions as examples from population
dynamics rather than as models of epidemics.

3.4.2 Immediate and instantaneous spore production, exponential growth of lesion density

In the simplest case, new spores are assumed to be
produced immediately after infection (latency period is
zero), they are all produced at the same instant
(infectious period is infinitesimatelly small) and
there is no exhaustion of non-infected sites. Of
course, in real epidemics spore movement is always fast
relative to lesion development, exactly the opposite of

our assumptions.

The number of new effective spores per deposited effective spore (a deposited effective spore is, under assumptions *A.3.ii.* and *A.3.vi*, equivalent to new lesion) will be denoted by q.

$$q = R\ I\ (1 - G) \tag{3.27}$$

where R is the number of spores produced per sporulating lesion per unit of time, I is the probability of infection and G is the fraction of spores removed from an epidemic (fallen on the ground, blown outside a field, etc.). The probability of infection, I, can be defined as follows:

Definition D.4. The probability of infection I equals the fraction of spores falling on vacant sites which turn these into infected sites.

Usually, it is not possible to measure I and G independently. The term $I\ (1 - G)$ is analogous to the effectiveness of a spore, E, as defined by Zadoks and Schein (1979). However, because I and G belong to completely different phenomena, they are introduced here explicitly. The fraction of successful spores, $E = I\ (1 - G)$, can be determined experimentally.

The parameter I comprises (is dependent on) three different phenomena: crop infectibility, pathogen infectivity, and suitability of the environment. It is a measure of the reaction of the crop to the specified pathogen in the specified environmental conditions. I varies from 0 to 1:

$I = 0$ - no spore will lead to a lesion,

$I = 1$ - all spores deposited on vacant sites will produce lesions,

$0 < I < 1$ - the crop is partially resistant (and thus also partially susceptible) and/or

the pathogen is partially virulent
and/or the environment is only partially
favourable.

This parameter is related to Barrett's (1980) fitness
of the pathogen on the particular host (under the
particular conditions).

The probability of infection, I, is one of the
parameters, in which vertical (I = 0) and horizontal (0
< I < 1) resistance (Van der Plank, 1963) of the crop
can be expressed. The parameter I can vary with space
to describe a non-uniform crop, a non-uniform
distribution of the pathogen, or non-uniform
environmental conditions, and with time to describe
crop resistance varying with plant development, time
diversity of pathogen behaviour, or time dependent
changes of environmental conditions.

The number of new effective spores per deposited
effective spore, q, is here taken to be constant during
the epidemic (exponential growth of the lesion
density).

Under the assumptions stated above, the production
term, denoted as P, is given by:

$$P = R \frac{\partial L}{\partial t} \quad [\mathrm{NL}^{-2}\mathrm{T}^{-1}] \tag{3.28}$$

where R is the number of spores produced by a
sporulating lesion, $\partial L/\partial t$ is the rate of change of the
lesion density L. This rate of change equals the rate
of spore deposition multiplied by the effectiveness:

$$\frac{\partial L}{\partial t} = E \delta S. \tag{3.29}$$

Substituting (3.29) into (3.28) with application of
(3.27) leads to:

$$P = q \delta S.$$

Substituting this in equation (3.26) finally gives:

$$\frac{\partial S}{\partial t} = D \left[\frac{\partial^2 S}{\partial x^2} + \frac{\partial^2 S}{\partial y^2} \right] - \delta S + q \delta S \qquad (3.30)$$

In Van der Plank's (1963) model, $(q-1) \delta$ would be the exponential growth rate (denoted by him as r_1). However, Van der Plank's equations deal with lesions rather than spores (lesions cannot disappear). Here, the disease is examined from a 'spore point of view', so that the term $(q-1)$ must be used instead of q (a deposited spore disappears, assumption *A.3.iv*). Note that Van der Plank considered a 'point' model (he did not take into account spatial aspects of disease) and <u>numbers</u> of lesions, whereas here S is interpreted as <u>density</u> of spores.

Inserting the new variable, $\beta = (q-1) \delta$, and setting $\vec{r} = (x,y)$, the solution of equation (3.30), for the case of an initial infection with one spore at the point $\vec{r} = \vec{0}$, is (Wyld, 1976; Morse and Feshbach, 1953):

$$S(\vec{r},t) = \frac{1}{4\pi Dt} \exp \left[\beta t - \frac{r^2}{4Dt} \right] \qquad (3.31)$$

If the focus was started by more spores, the right hand side of equation (3.31) should be multiplied by the number of these initial spores.

According to equation (3.31) the spore density $S(\vec{r},t)$ equals $\exp(\beta t)$ times the Gauss (normal) distribution with variance $\sigma^2 = 2Dt$ ($[L^2]$). The term $\exp(\beta t)$ describes the increase of spore density. Equation (3.31) describes the distribution of the density of spores still air-borne. Substitution of (3.31) into (3.29) gives the rate of change of the lesion density:

$$\frac{\partial L}{\partial t} = E\,\delta\,\frac{1}{4\pi Dt}\,\exp\left[\,\beta\,t - \frac{r^2}{4Dt}\,\right] \qquad (3.32)$$

The lesion density distribution function is the result of the Gaussian distribution of spores still in the air and of their deposition at a constant rate.

In the limit for large values of time, the distribution of the first generation lesions in the horizontal plane is described by the Bessel distribution (Broadbent and Kendall, 1953; Williams, 1961; Van den Bosch et al., 1988a, b). It resuls from diffusion and deposition of spores. When normalized, this distribution is called the `contact distribution' (see Van den Bosch et al., 1988a, b). The result in the field is seen as a focus.

To determine the velocity of focus expansion, additional definitions must be given. Ideally, the front of the focus is defined as the borderline which divides the plane into two different regions: (1) the region with disease and (2) the region free from disease. In the first region $L(\vec{r},t) > 0$, in the second one $L(\vec{r},t) = 0$. But, $L(\vec{r},t)$ as described by the `diffusion theory' is a continuous function, which for $t > 0$ is greater than zero everywhere, even though L soon becomes very small with increasing distance. Therefore, this definition would always place the position of the front at infinity. To handle this situation, an operational definition (see Zadoks and Schein, 1979, p. 10) must be adopted:

Definition D.5. The front of the disease focus is the borderline between the region where $L(\vec{r},t) > L_o$ and the region where $L(\vec{r},t) < L_o$, L_o being a small, positive number.

Now the position in space of the front of the expanding focus is the position where $L(\vec{r},t) = L_o$. This definition can be used in mathematical considerations as well as in practical field work (Buiel et al., 1989). In general, the front is not localized at a constant position throughout time, so that it is possible to determine the velocity of its movement.

Definition D.6. The velocity of focus expansion is the velocity of the displacement of its front ($d\vec{r}/dt$).

This velocity will be denoted $\vec{c}_o$. It is a vector (characterized by its value and its direction), but if a focus expands radially (the direction of the velocity $\vec{c}_o$ is the direction of $\vec{r}$) only the length of the velocity vector, c_o, must be determined.

The velocity of displacement of the spore cloud front (which is equivalent to the focus front), can be obtained by looking for the time dependence of the position of a constant spore density in the air. The appropriate method was given by Okubo (1980) (following Kendall, 1948). Setting $S(\vec{r},t)$ in equation (3.31) equal to ε, where ε is a small positive constant, and solving for r leads to:

$$\beta\, t - \frac{r^2}{4 \cdot D \cdot t} = \ln\,(4\,\pi\,D\,\varepsilon\,t)$$

After a few simple mathematical operations, we obtain for high values of time (terms less than proportional to t are disregarded):

$$r = 2\,\sqrt{D\,\beta\,t}$$

The velocity of focus expansion is <u>asymptotically constant</u>. The asymptotic displacement velocity of the front of a focus, $c_o = \lim\limits_{t \to \infty} dr/dt$, is found to be:

36

$$c_o = 2 \sqrt{D\,\beta}$$

After translation into original symbols the last equation becomes:

$$c_o = 2 \sqrt{D\,(q-1)\,\delta}\,, \qquad q \geq 1 \tag{3.33}$$

Note that necessarily $q \geq 1$; when absorption exceeds production the disease does not develop, $c_o = 0$, and the spore cloud disappears.

An identical result for the velocity of displacement of the spore cloud front can be obtained by integrating equation (3.31) over the region almost free from spores ($r > R_{max}$, where R_{max} is the radius of the region in which $S(\vec{r},t) > \varepsilon$), and putting this integral equal to a very low number. Almost no spores for $r > R_{max}$ means that $S(\vec{r},t) < \varepsilon$, where ε is a positive, very low number. This method was also used by Okubo (1980).

3.4.3 Immediate and instantaneous spore production, logistic growth of lesion density

Starting with the same assumptions about the latency and infectious periods as in Section 3.4.2 but considering logistic growth of the lesion density, we should multiply the right-hand side of equation (3.29) by $(1 - L/L_{max})$, where $L = L(\vec{r},t)$ is the lesion density, and L_{max} is the maximum possible lesion density. The new equation multiplied by the number of spores produced by a single lesion, R, again describes the time rate of spore production. Equations (3.26) and (3.29) take the following form:

$$\frac{\partial S}{\partial t} = D \left[\frac{\partial^2 S}{\partial x^2} + \frac{\partial^2 S}{\partial y^2} \right] - \delta\, S + R\, \frac{\partial L}{\partial t}$$

$$\frac{\partial L}{\partial t} = E\, \delta\, S \left[1 - \frac{L}{L_{max}} \right] \tag{3.34}$$

Near to the advancing front, where the lesion density is much lower than L_{max}, the Okubo (1980) method can be applied to find the velocity of displacement of the focal front described by equations (3.34). This method leads to the following inequality for the velocity of focus expansion, c:

$$c \geq c_o = 2 \sqrt{D\,(q-1)\,\delta} \tag{3.35}$$

c_o is the minimum speed at which a spore cloud front can move. Note that c_o is equal to the speed of displacement determined in Section 3.4.2, equation (3.33). Diekmann (1978, 1979) and Thieme (1977, 1979) proved by a more general approach that in the case of focal expansion from a localized infection, only this minimum speed is realized. For a nice heuristic argument why the minimal velocity is the only one that counts see Van den Bosch et al. (in prep.).

3.4.4 Latency period p, instantaneous production of spores, exponential growth of lesion density

In actual situations, the latency period cannot be neglected. We consider a model in which the infectious period is still taken to be only one instant and in which epidemic growth is supposed not to be limited by exhaustion of susceptible sites. Spores are produced by lesions, which were created by deposited spores, one latency period before. Therefore, the production term

will refer to the density of spores, deposited one
latency period before. The rate of change of the spore
density is described by equation (3.26), but the
production term is now:

$$P = R \; \frac{\partial L(\vec{r}, t-p)}{\partial t} \tag{3.36}$$

where p is the latency period. Therefore, equation
(3.26), after substitution of (3.36) with application
of (3.29), takes the following form:

$$\frac{\partial S(\vec{r}, t)}{\partial t} = D \left[\frac{\partial^2 S(\vec{r}, t)}{\partial x^2} + \frac{\partial^2 S(\vec{r}, t)}{\partial y^2} \right] -$$

$$\delta \; S(\vec{r}, t) + \eta \; S(\vec{r}, t-p) \tag{3.37}$$

where p is the latency period, and $\eta = q \, \delta$.

The most interesting result of this model is the
velocity of focus expansion, c. This velocity can be
considered within the framework of the theory developed
by Diekmann and Thieme. They proved generally that (as
in Section 3.4.3) there exists a value c_o such that $c \geq$
c_o. The minimum speed of focus expansion is implicitly
defined by a pair of equations in λ_o and c_o:

$$f_{c_o}(\lambda_o) = D \lambda_o^2 - c_o \lambda_o - \eta \; e^{c_o p \lambda_o} = 0$$

$$\frac{df_{c_o}}{d\lambda}(\lambda_o) = 2 D \lambda_o - c_o - \eta \; c_o \; p \; e^{c_o p \lambda_o} = 0 \tag{3.38}$$

Again, in the case of focus expansion from a localized
infection, only the minimum speed c_o is realized.

3.4.5 Latency period *p*, infectious period *i*, exponential growth of lesion density

If both the infectious and latency periods are not negligibly small, and the spore population grows exponentially, the rate of change of the lesion density takes the form of equation (3.29) again, but the production term in (3.26) is:

$$P(r,t') = \int_{-\infty}^{t'} K(t' - \tau) \, \frac{\partial L(\vec{r},\tau)}{\partial t} \, d\tau$$

where:

$K(t-\tau)$ is the function describing the time dependence of the number of daughter spores produced at time *t* per unit of time by a mother lesion, which was created at time *t-τ*,

$\partial L(\vec{r},\tau)/\partial t$ is the rate of lesion production at $\vec{r}$ and *τ*, which is proportional to the rate of spore deposition at the same place and time; $\partial L(\vec{r},\tau)/\partial t = E \, \delta \, S(\vec{r},\tau)$.

An equivalent, sometimes more convenient form of the production term is:

$$P = \int_{0}^{\infty} K(\tau) \, \frac{\partial L(\vec{r}, t-\tau)}{\partial t} \, d\tau \tag{3.39}$$

The diffusion equation (3.26) takes the form:

$$\frac{\partial S(\vec{r},t)}{\partial t} = D \left[\frac{\partial^2 S(\vec{r},t)}{\partial x^2} + \frac{\partial^2 S(\vec{r},t)}{\partial y^2} \right] - \delta \, S(\vec{r},t) +$$

$$\int_{0}^{\infty} K(\tau) \, \frac{\partial L(\vec{r}, t-\tau)}{\partial t} \, d\tau \tag{3.40}$$

Again, this is a special case of the Diekmann-Thieme theory.

Looking for the velocity of focus expansion, c, a method analogous to the one of Section 3.4.4 can be used. The analog of the system of equations (3.38) is:

$$\frac{df_{c_0}}{d\lambda}(\lambda_0) = 0$$

$$f_{c_0}(\lambda_0) = 0 \tag{3.41}$$

where

$$f_c(\lambda) = D\lambda^2 - c\lambda - \delta + \tilde{K}(c\lambda) = 0$$

with

$$\tilde{K}(p) = \int e^{-p\tau} K(\tau)\, d\tau$$

the so-called Laplace transform of the spore production kernel.

As in Section 3.4.4, it is possible to prove within the framework of the Diekmann-Thieme theory, that the velocity of focus expansion $c \geq c_0$ obtained from (3.41), and that for the case of a localized initial infection only this minimal velocity c_0 is realized.

3.4.6 The general case

In a realistic approach, the latency and infectious periods cannot be neglected, and the growth of the lesion density is bounded by a maximum value (the density of sites). In the usual point model (the model which does not take into account the spatial development of the disease), the development in time is described by Van der Plank's equation (Van der Plank, 1963, p. 100) here rendered as:

$$\frac{dL(t)}{dt} = R_c \left[L(t-p) - L(t-p-i) \right] \left(1 - \frac{L(t)}{L_{max}} \right)$$

$$(3.42)$$

where:

$L(t)$ - the number of lesions at time t,

R_c - the number of daughter lesions per sporulating mother lesion per unit of time = the basic infection rate corrected for removals,

p - the latency period,

i - the infectious period,

L_{max} - the maximum number of lesions.

Note that in this case L and L_{max} are numbers and not densities. Equation (3.42) cannot be used here because the diffusion equation deals with spores and not with lesions. Therefore, we will try to find an equation analogous to (3.42), which describes the development of an epidemic in time, and which takes into account that lesions are produced by spores, whose distribution is described by the diffusion equation (3.26).

Not every spore deposited on a vacant site will change it to the infected state (Section 3.4.2.). The rate of production of new lesions at point $\vec{r}$ and time t is equal to the spore deposition rate $[NL^{-2}T^{-1}]$ on non-infected sites of leaves, $f(\vec{r},t)$, multiplied by the probability of infection I.

$$\frac{\partial L(\vec{r},t)}{\partial t} = I\, f(\vec{r},t) \qquad (3.43)$$

The right hand side of this equation is the deposition rate of effective spores, spores which will produce new lesions.

The rate of spore deposition, $[\partial S(\vec{r},t)/\partial t]_a$, at $\vec{r}$ and t is stated by equation (3.26). This rate should be corrected for removal of spores from the epidemic (spores that are dead, fall on the soil, etc.) and for

spores which fall on infected sites (and cannot infect
them again, assumption *A.2*). The deposition rate of
spores which can produce new lesions is:

$$f(\vec{r}, t) = (1 - G) \, \delta \, S(\vec{r}, t) \left[1 - \frac{L(\vec{r}, t)}{L_{max}} \right] \qquad (3.44)$$

where G is the fraction of spores removed from the
epidemic, and the term $(1 - L/L_{max})$ is the correction
factor for multiple infection (the fraction of vacant
sites, as in Van der Plank's equation), where now, the
number of lesions depends on a point in space $\vec{r}$.

Substituting equation (3.44) into (3.43), the
deposition rate of effective spores which will produce
new lesions is obtained.

$$\frac{\partial L(\vec{r}, t)}{\partial t} = E \, \delta \, S(\vec{r}, t) \left[1 - \frac{L(\vec{r}, t)}{L_{max}} \right] \qquad (3.45)$$

where $E = I (1-G)$ is the effectiveness (Zadoks and
Schein, 1979).

Because the new spores are produced by lesions, the
production term of the diffusion equation takes the
form (3.39). Substitution of (3.39) into equation
(3.26) gives:

$$\frac{\partial S(\vec{r}, t)}{\partial t} = D \left[\frac{\partial^2}{\partial x^2} + \frac{\partial^2}{\partial y^2} \right] S(\vec{r}, t) - \delta \, S(\vec{r}, t) +$$

$$\int_0^\infty K(\tau) \, \frac{\partial L(\vec{r}, t-\tau)}{\partial t} \, d\tau \qquad (3.46)$$

Equations (3.45) and (3.46) constitute the system of
partial differential equations, for spores and lesions
respectively, which is the mathematical formulation of
the 'diffusion theory' for focus development in time

and space.

In the most simple case $K(t-\tau)$ is a block function:

$$K(\tau) = \begin{cases} R & \text{for } p \le \tau \le p+i \\ 0 & \text{for } \tau < p \text{ or } \tau > p+i \end{cases} \tag{3.47}$$

where R is the number of spores produced by one sporulating lesion per unit of time. After substitution of (3.47) in (3.39) and calculation of the integral, the source term is written as:

$$P = R \left[L(\vec{r}, t-p) - L(\vec{r}, t-p-i) \right] \tag{3.48}$$

Substituting (3.48) in equation (3.26), the following diffusion equation can be written:

$$\frac{\partial S(\vec{r}, t)}{\partial t} = D \left[\frac{\partial^2}{\partial x^2} + \frac{\partial^2}{\partial y^2} \right] S(\vec{r}, t) - \delta\, S(\vec{r}, t) +$$

$$R \left[L(\vec{r}, t-p) - L(\vec{r}, t-p-i) \right] \tag{3.49}$$

This form of the diffusion equation together with equation (3.45) constitute a system of partial differential equations analogous to (3.45), (3.46).

The asymptotic velocity of focus expansion is one of the results, which can be obtained from the system (3.45), (3.46) by analytical methods. Van den Bosch et al. (1988a, b) discuss these results in detail. In other cases of practical importance, the system (3.45), (3.46) is too complex for an analytical solution. A numerical solution can be obtained by means of the computer package PODESS (Partial and/or Ordinary Differential Equations Systems Solver), written to this purpose (Appendix A). Some of its results will be discussed in the following chapters.

3.5 CONSIDERATIONS FOR THE APPLICATION
OF THE DIFFUSION MODEL

3.5.1 A guide-line for the application
of the diffusion model

The limitations of the theory presented in the
Sections 3.3 and 3.4 should be remembered and carefully
considered before applying the theory.

The diffusion equation is only exactly valid in the
limit for: (1) a large velocity of spores, (2) a small
mean free path for scattering, and (3) a large mean
free path for absorption. These quantities should tend
to their respective limits in such a way that $D = v
\lambda_s/2 = constant$, and $\delta = v/\lambda_a = constant$. Of course,
the limit is not realized for real plant disease foci.
Therefore, the guide-line in application should be the
following:

*Use the diffusion equation to describe focus
development, if*

1. *the mean free path for scattering (`mixing
   length') is much shorter than the distance
   travelled by a spore during the time-period of
   interest,*
2. *the mean free path for absorption is much higher
   than the mean free path for scattering.*

Actually, the mean free path for absorption should be
of the order of magnitude of the spore velocity
multiplied by the time-period of interest. The diameter
of the `solution' region must be considerably smaller
than the distance travelled by a spore along its
trajectory during the time-period of interest.

The approach of the `diffusion theory' describes
only those processes which are continuous in time and
space. Therefore, the theory is applicable to focus
development, but it is not applicable to stochastic

processes which deal with low numbers of spores or lesions and which also are a part of an epidemic. For instance, the description of daughter focus formation will require a different, stochastic approach. This approach will be discussed in Section 8.5, some of its results will be presented in Chapter 9.

3.5.2 Restrictions to be imposed on parameter values

The diffusion equation (3.46) was derived with assumptions which limit the `diffusion theory' parameter values.

The constancy, over space of the value of the diffusion coefficient restricts the field size to values low enough to neglect spatial variation of D. In the simulated field, the crop should be uniform.

Theoretically, the diffusion equation describes situations when spore movement is completely at random on an infinitesimally small scale. In reality, it should be applied to situations when the mixing of the spores is strong enough to assume that, within one time-step of integration of the numerical solution, the direction of a spore's movement can be changed to another independent direction.

The derivation of the diffusion equation used an approximation of the spore flux by the first order Taylor expansion terms. Therefore, the rates of production and deposition must be low enough to keep the higher order terms negligibly small.

The restrictions stated above do not indicate the exact limits of the parameter ranges allowed by the diffusion approximation. These limits depend on a particular application and on the required degree of realism of the `diffusion model'. Thus, the limits must be set separately for every application. The following section discusses this problem using a few real-life examples.

3.5.3 Real parameter values

In the case of focus development of an airborne
plant disease, where spore dispersal takes place inside
the crop canopy, the above restrictions usually are of
no great consequence. The mean free path for scattering
inside a crop (also called `mixing length') varies from
0.016 m for grass to 0.23 m for maize (Goudriaan, 1977,
p. 112). The velocity of spores is the wind speed,
according to Chamberlain (1967, p. 140). He showed,
that the relaxation time - "...the time for the
particle to accommodate itself to the motion of
surrounding air..." - is a few milliseconds for large
spores; as it is proportional to the square of the
spore radius, the relaxation time for smaller spores is
even shorter. For light and moderate winds, wind speeds
vary from 0.4 to 2.6 m/s at 1 m above ground level
(Chamberlain, 1967, p. 149). Inside a crop, wind speed
is lower but, excluding the layer just above the soil
surface, it is usually in the order of tens of
centimetres per second (McCartney and Fitt, 1985, pp.
118 - 119). During tens of seconds spores can travel
distances much longer than the mean free path for
scattering. This is equivalent to strong mixing of
air-borne spores during such periods: "... at least
two-thirds of the eddying energy is associated with
eddies of less than 5 seconds..." (P.H. Gregory, 1973,
p. 73).

The requirement of a low deposition of spores poses
a problem. Low deposition is indeed the case with
Puccinia polysora in maize, examined by Cammack (1958;
also Van der Plank, 1963, p. 282, and Gregory, 1968).
His data show a decrease of the average number of
pustules per plant with distance from the point of
initial inoculation. The curve representing the primary
gradient (10 days after inoculation) allows to assess
the parameters of the Bessel distribution which in this

case describes the pustule distribution. The variance of the Bessel function, which equals $D/\delta = \lambda_s \lambda_a$, (Broadbent and Kendall, 1953; Van den Bosch et al., 1988a, b) is estimated to be a few meters (the method of calculation is given in detail by Williams, 1961). Together with Goudriaan's (1977) statement that his `mixing length' (which is equivalent to our mean free path for scattering λ_s) is in the order of magnitude of centimetres to decimetres, this value of variance suggests a high value of the mean free path for absorption in comparison to the value of the mean free path for scattering. The direction of a spore's motion can be changed many times before the spore is deposited. Thus a spore is deposited at site `chosen' at random. The mixing process is far more intensive than the deposition process (the rate of changing a direction of movement is high compared to the deposition rate).

Not always is the situation so nice. In the case of stripe rust (*Puccinia striiformis*) on wheat Zadoks (1961, p. 102) stated "The first-generation focus consists of one infected leaf only, the second-generation focus counts up to ten leaves and covers a drill length of 10 cm.". In this case, the mean free path for absorption is in the order of magnitude of centimetres, about equal to the the mean free path for scattering. The place of a spore's deposition is not independent from its original direction of motion immediately after take-off. Thus, the `diffusion theory' is no longer valid. In the case of stripe rust, described above, spores were dispersed by rubbing and splash mechanisms rather than by turbulent diffusion. The example shows why the `diffusion theory' can be used only for analysis of air-borne plant diseases.

The mean free path for absorption depends strongly on crop density. The numerical analysis of Tyldesley

(1967, p. 28) for a crop-free region showed that for particles of 10 μm diameter the fraction still airborne at 100 m from the source varies from 0.90 to 0.98, depending on the model of spore deposition used. In such a situation the mean free path for absorption is in the order of magnitude of hundreds of meters. Thus assuming the size of eddies in the air above crop layer to be of the order of magnitude of 1 m, the diameter of the region of solution can be hundreds of meters. Such a large size of the solution region allows to solve the problem of focus expansion in the range of hundreds of meters by spore dispersal above a crop. This point is taken up again in Section 8.4 and in Chapter 9 (multiple dispersal mechanism).

3.5.4 Technical aspects of the numerical solution

The time period of interest for phytopathologists is an hour or a day. But, in the case of a numerical solution of the system of equations (3.45) and (3.46), the period of interest is the time-step of integration. Usually, a system of `diffusion theory' equations is solved by a method with a self-adapting time-step of integration, whose value is chosen so as to keep the error of numerical integration at a low level (specified by the user). This requires the time-step of the numerical integration to be of the order of magnitude of the shortest time constant of the system. If the time-step of numerical integration is a few minutes only, the hour or day of phytopathological interest is obtained by solving the system of equations during as many time-steps as is necessary to complete that hour or day.

In the runs of the `diffusion model' the space representing a field is finite. Therefore, some spores travel to the space's boundary, where their further story is `decided' by the boundary conditions imposed

on the solution of the diffusion equation. There are
three possibilities: (1) Dirichlet conditions (boundary
conditions specify the function), (2) Neumann
conditions (boundary conditions specify the normal
derivative, and (3) mixed conditions (Ames, 1977). The
choice of the boundary conditions depends on the
situation to be simulated. In our simulations, the
boundary conditions were specified by equating the
second normal derivative at the boundary point to the
one at the nearest grid point in the direction normal
to the boundary. When the initial infection is placed
at the centre of the field, this condition means that
the values of the spore and lesion densities at the
boundary and at the nearest point are equal. The
influence of this boundary condition, which is of
Dirichlet type, on the result of the numerical solution
of the 'diffusion model' will by studied in Section 5.5
by means of sensitivity analysis.

4 VALIDATION OF THE 'DIFFUSION THEORY' IN THE HORIZONTAL PLANE

4.1 INTRODUCTION

Plant disease development in space and time can be treated by a variety of methods: deterministic computer simulation (Zadoks and Kampmeijer, 1977; Kiyosawa, 1976), stochastic computer simulation (Minogue and Fry, 1983), analytical treatment of integro-differential equations (Kermack and McKendrick, 1927; Diekmann, 1978, 1979; Thieme, 1977, 1979; Van den Bosch et al., 1988a, b, c) or the 'diffusion theory' (Chapter 3). The technical aspects of these methods may be very different, but their results should be consistent; they have to reflect the nature of the process described.

This chapter compares the results of some computer simulation models and some experimental data to the results obtained by numerical solution of a system of partial differential equations (Chapter 3).

4.1.1 Parametrization

The parameters required by the 'diffusion model' belong to distinct groups. The elements within each group are related by similarities in their meaning for the theory and in their method of measurement. Sometimes they cannot be measured separately from other parameters belonging to the same group.

1. Spore production parameters:
   a. E - effectiveness - is the proportion of spores produced which after deposition on healthy plant tissue will produce lesions.
   b. R - reproductivity - is the number of spores produced per sporulating lesion per day.
   c. p - latency period - is the time in days from the

> deposition of a successful spore until the start of spore production by the ensuing lesion,
>
> d. i - infectious period - is the period in days during which a lesion produces spores.

Choice of these parameters assumes that the reproductivity can be described by a block function, with its non-zero value between two time points: the beginning and the end of the infectious period of a lesion. If this is not the case, R, p and i should be replaced by a function describing the time dependency of spore production by a lesion.

Often, it is not possible to measure R and E separately. When spore production can be described by a block function, the number of daughter lesions per mother lesion per day can be used i.e. the infection efficiency, E, multiplied by the reproductivity, R. In the case of another time dependency of the spore production function, time dependent function for E and R or for $E \cdot R$ should be determined experimentally.

2. Spore movement parameter:

   a. D - diffusion coefficient.

In the scattering model (Section 3.3.4)

$$D = \lambda_s \, v \, / \, 2$$

where λ_s is the mean free path for scattering (analogous to the mixing length (Goudriaan, 1977)), and v is the spore velocity.

3. Spore 'survival' parameter:

   a. δ - the rate of spore deposition.

In the scattering model

$$\delta = v \, / \, \lambda_a$$

where v is the spore velocity and λ_a is the mean free path for absorption.

Sometimes, D and δ are experimental results, which can be used directly in a simulation run, but it is difficult to measure these parameters separately. Experimental determination of the contact distribution (sensu Van den Bosch et al., 1988a, b, c) gives the ratio, D/δ (Williams, 1961).

4. `Crop' parameters:
   a. number of available sites,
   b. spatial distribution of a crop and its variation with time.

The values of all parameters described in Section 4.1.1. are subject to regular or stochastic variation. The use of constant values is adequate for a preliminary analysis, but a more detailed analysis requires determination of changes of these parameter values with time and/or space.

4.2 COMPARISON WITH OTHER MODELS

4.2.1 Minogue and Fry's model
– one spatial dimension and time

Minogue and Fry modelled disease development in time and one-dimensional discretized space. In their model, sporulation and spore dispersal are stochastic processes. At low population density, the total number of offspring produced per parent lesion during its lifetime, n, has a Poisson distribution:

$$h(n) = \alpha^n \exp(-\alpha) \: / \: n!$$

where α is the mean number of daughter lesions per parent lesion. Daughter lesions are produced by a mother lesion of age p till $p+i$, where p is the latency period and i is the infectious period. The distribution of times at which daughter lesions occur is a block function:

$$
z(t) = \begin{cases} i^{-1}, & \text{for } t_o + p \le t \le t_o + p + i \\ 0, & \text{for } t < t_o + p \text{ or } t > t_o + p + i \end{cases}
$$

where t_o is the time at which a mother lesion was initialized. Assuming that spores moved straight away from this point of origin, the only mechanism leading to a decrease of the density of airborne spores being deposition with probability a at each crossing of a space cell, Minogue and Fry took the distribution of deposited spores to be a double geometric one:

$$
f(x) = [a / (2 - a)] (1 - a)^{|x|} \tag{4.1}
$$

where $|x|$ is the absolute value of the distance from the plant of origin, and $f(x)$ is the probability that a spore will travel that distance before landing. The probability of infection, conditional on a spore being deposited, of the jth plant was assumed to be proportional to the noninfected proportion of its tissue:

$$
Q_j = 1 - (y_j / K)
$$

where y_j is the number of lesions on plant j and K is the maximum number of lesions that can occur on a jth plant.

Gradients of the lesion distribution in a field, the displacement velocity of the disease front, and their dependence on the values of the parameters were examined. The spore distribution function (4.1), arbitrarily chosen by Minogue and Fry, happens to correspond to the one derived on theoretical grounds for the decay of spores with distance due to diffusion and eventual deposition, by Williams (1961) and Broadbent and Kendall (1953) (see also Van den Bosch et al., 1988b). However this correspondence is not

54

immediate, because Minogue and Fry disregarded the influence of diffusion on the spore density distribution. Luckily, this difference between the two models compared can be overcome, because the double geometric distribution can be directly translated to the double exponential one (i.e. the marginal distribution resulting from the `diffusion theory').

A Parameters

Minogue and Fry used several parameters in their simulations. Some were estimates of `disease parameters', others were parameters of the functions which govern sporulation of lesions and dispersal of spores. The `diffusion model' uses parameters which are not always consistent with the parameters of Minogue and Fry. So, some parameters were used without change, some were reinterpreted, and others had to be translated.

1. Unchanged parameters.
   - p - latency period   [T]
   - i - infectious period   [T]
   - L_{max} - the maximum number of lesions per
          unit of length $[NL^{-1}]$

Ad L_{max}. Minogue and Fry's K (the maximum number of lesions that can occur on a plant) is equivalent to the maximum lesion density L_{max}, as a plant occupies a unit of length.

2. Reinterpreted parameters.
   - R - number of spores produced by a
          single sporulating lesion per unit
          of time $[NN^{-1}T^{-1}]$
   - E - infection efficiency   [1]

Ad R, E. Minogue and Fry use M - the mean number of offspring produced per infectious lesion per unit of time at low population densities. It almost equals $R \cdot E$ of the `diffusion theory' or R_c defined by Van der Plank (1963). The only difference is that in Minogue and Fry's model the realized number of offspring is a Poisson random variable, whereas we assume that this random variable can be safely replaced by its mean, as we are always dealing with large numbers of sites.

3. Translated parameters.
- D - diffusion coefficient $[L^2 T^{-1}]$
- δ - rate of spore deposition $[T^{-1}]$

Minogue and Fry used σ^2 (the variance of the spore dispersal function) as the parameter measuring the distribution of daughter lesions. They assumed that spores move straight away from their point of origin and land on a plant with probability a, which is constant for all plants, and that their dispersal in either direction from the source plant is equally likely. Minogue and Fry derived that the variance of the resulting double geometric distribution function equals

$$\sigma^2 = 2 \ (1 - a) \ / \ a^2 \tag{4.2}$$

Before trying to relate Minogue and Fry's a to our D and δ it should be noted that, in Minogue and Fry's view, spore dispersal is instantaneous. For the 'diffusion theory' it corresponds to both D and δ being infinite, but they tend to infinity in such a way that D/δ takes a finite value. In the case of one-dimensional diffusion and deposition, the distribution of spores not yet deposited can be derived in the same manner as (3.31) was derived for two-dimensional space. The result is:

$$S(x,t) = \frac{1}{\sqrt{4\pi Dt}} \ \exp \left[-\delta \ t - \frac{x^2}{4Dt} \right]$$

The number of spores landing at a distance x from the

source is the integral of $S(x,t)$ with time, t, calculated from 0 to infinity. The result is exactly equal to the one-dimensional marginal distribution of spores deposited in two-dimensional space (see Van den Bosch et al., 1988b):

$$\tilde{S}(x) = 1/2 \sqrt{\delta/D} \exp\left[-\sqrt{\delta/D}\,|x| \right] \qquad (4.3)$$

Because a fraction of the deposited spores E initialize lesions, equation (4.3) multiplied by this correction factor describes the lesion distribution at low lesion densities (when the fraction of tissue already infected has little influence).

Comparing formulas (4.1) and (4.3), the approximate values of D/δ corresponding to Minogue and Fry's values of σ can be calculated:

$$D/\delta = \left[\ln (1 - a) \right]^{-2} \qquad (4.4)$$

where a can be calculated from (4.2) with the values of σ chosen by Minogue and Fry.

B Results

Results were obtained by running the programme PODESS (Appendix A) on a VAX 8600. The solution interval (a unit of 'solution' time) was one day. As in Minogue and Fry's model, the field was one-dimensional (one line of crop). Its length was 40 units (a unit of length is a distance occupied by a single plant).
The parameters common to all runs were:
1. L_{max} = 50. - maximum lesion density (per plant) $[NL^{-1}]$
2. D = 5. - diffusion coefficient $[L^2T^{-1}]$
3. E = 1. - infection efficiency $[1]$

Ad 3. The value of D was chosen to keep λ_s always lower than that λ_a, where in the scattering model $D = \lambda_s \cdot v/2$ and $\delta = v/\lambda_a$; only $D/\delta = \lambda_s \cdot \lambda_a$ can be calculated from (4.4).

The other parameters (varied in different runs) were:

6. R – number of spores produced by a single
           sporulating lesion per day $[NN^{-1}T^{-1}]$
7. p – latency period (in days) $[T]$
8. i – infectious period (in days) $[T]$
4. δ – rate of spore deposition $[T^{-1}]$.

The initial inoculation, by a single spore, occurred at the left end of the field (point 0.).

Four runs for different parameter values (Table 4.1), were performed (Fig. 4.1). Three additional runs were made for three values of the rate of spore deposition (δ = 25.6, 2.6, 1.3), keeping other parameters values as for run 1.

The curves in figures 1A to 1D of Minogue and Fry show simulated populations of lesions as functions of the distance from the point of origin. These curves

Table 4.1. A comparison of the 'diffusion model' and the model by Minogue and Fry. Values of the input parameters for the first four runs of the 'diffusion model'. R - number of offspring per sporulating lesion per day, p - latency period, i - infectious period, δ - rate of spore deposition.

Par.	Run 1	Run 2	Run 3	Run 4
R	0.5	1.0	0.5	0.5
p	3.0	3.0	3.0	7.0
i	5.0	5.0	10.0	5.0
δ	4.8	4.8	4.8	4.8

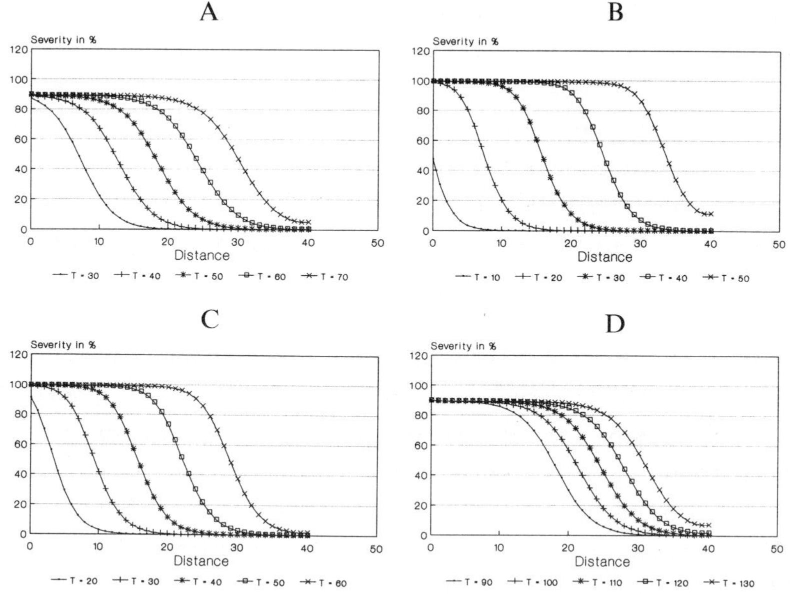

Fig. 4.1. Lesion density as a function of distance from the point of initial inoculation for various time instants after inoculation. The X-axis shows distance in units of length (plants), the Y-axis disease severity in percent of the maximum number of lesions. Parameter values are given in Table 4.1 (compare to Fig. 1 in Minogue and Fry, 1983a). **A,** results of run 1. **B,** results of run 2. **C,** results of run 3. **D,** results of run 4.

were compared with the printouts of runs 1 to 4. The gradient values obtained by means of the 'diffusion model' are shown in Table 4.2. The only difference is in the times at which similar curves of lesion density versus distance appear. Curves produced by the 'diffusion model' appear earlier than those produced by Minogue and Fry's calculations. The initial phase of focus build-up in the 'diffusion' model is shorter than

in the Minogue and Fry model. The difference seems to
due to the deterministic nature of the `diffusion
theory' as compared to the stochastic one of the
Minogue and Fry model. Minogue and Fry assumed the
number of offspring produced by a sporulating lesion to
be a Poisson distributed random variable. Therefore, a
low rate of deposition is usually translated into no

Table 4.2. A comparison of the `diffusion model' and
the model by Minogue and Fry. Values of the gradients
(assessed for 50% of the maximum disease severity) for
Fig. 1A - 1D of Minogue and Fry's model and runs 1 - 4
of the `diffusion model'. Gradients are expressed as
differences in the `percent of maximum number of
lesions per plant' at the two points nearest to 50%
severity (these points are one unit of length apart).
The runs of the `diffusion model' were performed with
values of parameters as in Table 4.1 and in text.

`Diffusion model' - gradients constant - model is deterministic	Minogue and Fry's model - gradients are variable - model is stochastic
Run 1 -8.0	Fig. 1A -3.8 to -7.9
Run 2 -11.9	Fig 1B -10.0 to -15.0
Run 3 -11.3	Fig. 1C -10.4 to -11.2
Run 4 - 7.6	Fig. 1D -7.0 to -7.5

60

lesion being initialized, which corresponds to an effective cut off of low lesion densities (of course there are also random jumps forwards, but these are rare, cut off being the usual pattern), thus slowing down the initial phase of the epidemic.

The results show good qualitative and quantitative consistency. The gradients for runs 2 and 3 (Table 4.2) are much steeper than for runs 1 and 4; this reflects the dependence of the gradient on the number of spores produced per infectious lesion per unit of time, R (compare results of the runs 1 and 2), and on the duration of the infectious period, i (compare results of the runs 1 and 3). Both models are consistent in that the latency period p has little influence on the gradient (compare results of the runs 1 and 4).

Runs 1, 5, and 7 were performed with values for δ = 4.8, 25.6, and 1.3, respectively. The values of the gradients for these runs and these of Minogue and Fry

Table 4.3. A comparison of the 'diffusion model' and the model by Minogue and Fry. Gradients for different values of the 'dispersion' parameter. Gradients are expressed as differences in the 'percent of maximum number of lesions per plant' at the two points nearest to 50% severity (these points are one unit of length apart). Symbols are explained in the text.

'Diffusion model'		Minogue and Fry's model	
δ	gradient	σ	gradient
25.6	-17.8	0.79	-18.0
4.8	-8.0	2.00	-8.0
1.3	-4.6	3.94	-3.5

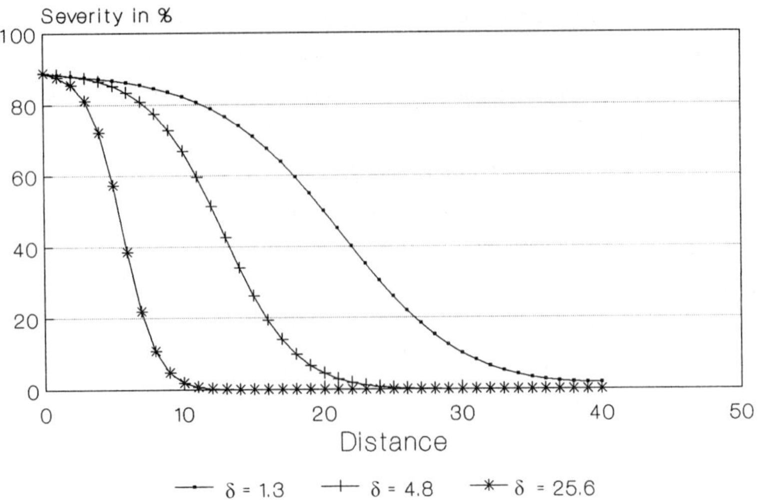

Fig. 4.2. Lesion density as a function of distance from
the point of initial inoculation for various values of
the rate of spore deposition, δ, (25.6,   4.8,   1.3)
produced by runs 5, 1, 7. The X-axis shows the mean
free path for absorption in units of length (plants),
the Y-axis is disease severity in number of lesions.
Other parameter values are given in Table 4.1 (compare
to Fig. 2 in Minogue and Fry, 1983a).

are shown in Table 4.3 (compare Fig. 2 of Minogue and
Fry to Fig. 4.2 of the present study). Again, the
results of the two models are qualitatively and
quantitatively consistent. This result is not a
surprise. In the limiting case of infinitely fast spore
dispersal, the 'diffusion model' contains only one
parameter with dimension length, $\sqrt{D/\delta}$, i.e. the nature
of the solution becomes almost independent of the
separate parameters D and δ as long as D/δ is kept
constant.

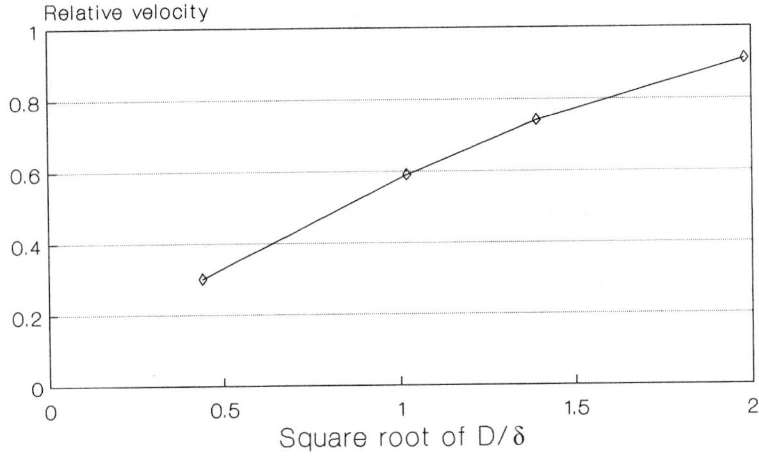

Fig. 4.3. Velocity of focus expansion as a function of the logarithm of the mean free path for absorption. Results for δ = 25.6, 4.8, 2.6, 1.3 are produced by runs 5, 1, 6, 7. The X-axis is the square root of D/δ, the Y-axis represents the velocity in units of length (plants) per 'solution' interval. Other parameter values are given in Table 4.1 (compare to Fig. 3 in Minogue and Fry, 1983a).

Another important parameter, which characterizes conquest of space by the disease, is the velocity of travel of the epidemic wave (or of a low constant value of the severity). The value of this velocity depends on the value of the 'dispersion' parameter. Fig. 3 and Table 1 of Minogue and Fry's paper present their results. Runs 1, 5, 6, and 7 show the results of the numerical approach by means of the 'diffusion model'. These results are compared in Table 4.4 (compare also

Fig. 3. of Minogue and Fry and our Fig. 4.3). As can be expected from the theoretical study of the case with infinitely fast spore dispersal, both models predict linear growth of the wave velocity with increase of the value of the `dispersion' parameter, variance for the Minogue and Fry model and $\sqrt{D/\delta}$ for the `diffusion model'. The left side of Table 4.4 also shows the increase of the velocity of focal expansion with time until reaching a constant value.

Table 4.4. A comparison of the `diffusion model' and the model by Minogue and Fry. Traveling wave velocity as a function of the `dispersion' parameter λ_a of the `diffusion' theory or σ of the Minogue and Fry's model.
† 1 - mean velocity between points at 0. and 10. units from the point of initial inoculation.
2 - mean velocity between points at 10. and 20. units from the point of initial inoculation.
3 - mean velocity between points at 20. and 30. units from the point of initial inoculation.
‡ 1 - velocity resulting from Minogue and Fry's model.
2 - velocity as assessed from Minogue and Fry's data (their Fig. 3) by linear regression.

`Diffusion model'				Minogue and Fry's model		
	velocity[†]				velocity[‡]	
δ	1	2	3	σ	1	2
25.6	0.26	0.30	–	0.79	0.22	0.20
4.8	0.51	0.57	0.59	2.00	0.51	0.50
2.6	0.64	0.70	0.74	2.74	0.57	0.67
1.3	0.81	0.87	0.91	3.94	1.07	0.95

C Discussion

The numerical analysis of the system of partial differential equations, which form the basis of the `diffusion theory', presents the gradient and the expansion velocity of the focus as functions of a set of parameters. High qualitative and quantitative consistency with the results of Minogue and Fry was obtained. However, it is important to notice that this consistency resulted from a fortunate consistency of the distribution function assumed by Minogue and Fry and the one resulting from the `diffusion theory' rather than from consistency in assumptions on spore dispersal mechanisms. Minogue and Fry assumed that only deposition is `responsible' for the decrease of the density of deposited spores with the distance from the plant of origin. The `diffusion theory' assumes that the spore distribution is the result of two processes: (1) turbulent diffusion, and (2) deposition.

Quite opposite to the results of the two models compared above is the opinion about the functional dependence of the gradient on the infection rate advocated by Vanderplank (1975). He writes (page 141): "From any given level of disease in an established epidemic, the gradient will be flatter as the infection rate is faster, other things being equal." This opinion is based on his equation (4.3) (Vanderplank, 1975 p. 105) for a nonspatial model, which was derived with the assumption: "When there are no waves and the epidemic is proceeding at a steady rate and y (here the fraction of infected host tissue) is relatively small (and definitely not exceeding 0.15), R can be estimated from r by the equation

$$r = R \left[e^{-p \cdot r} - e^{-(i+p) \cdot r} \right] ."$$

Vanderplank derived his equation (4.3) from a `point'
model (spatial development of the disease was
completely neglected). Such an equation cannot be the
base for `spatial' results (a gradient is a `spatial'
phenomenon). Morever, the gradient should be measured
`on the wave', because for a focus the wave is the
front of the epidemic.

The argument can be visualized by the following
example. At the level of applicability of Vanderplank's
equation (4.3) ($y < 0.15$), disease develops almost
exponentially. Assuming his `point' model at two
points, P_1 and P_2, with different disease levels, $y_1(0)$
and $y_2(0)$ (for $t = 0$), these levels will grow
exponentially with time, and so will grow their
difference. The growth will be proportional to the
exponent of the infection rate. Thus a higher value of
the infection rate will result in a higher value of the
difference between levels $y_1(t)$ and $y_2(t)$ (where $t >
0$). Therefore, the gradient will be steeper for the
higher than for lower infection rate (the gradient is
the first derivative with respect to the space
variable, so it is the limit of the difference between
the disease levels divided by the difference between
the positions of points P_1 and P_2, when the latter
difference tends to 0.).

Minogue and Fry interpreted the data in a paper by
MacKenzie (1976) as if slow rusting cultivars (low
infection rate) are characterized by steeper disease
gradients. A detailed inspection of the paper does not
confirm Minogue and Fry's interpretation. Table 1 and
Fig. 3 of MacKenzie's paper, which give the gradients
of a slow rusting wheat variety (Bonza 55) and of two
susceptible varieties (Pitic 62 and Penjamo 62), show
that the gradients of Bonza 55 are not significantly
different from the gradients of Pitic 62 and Penjamo
62. In addition, D.R. MacKenzie states: "Significant
differences in the regression slopes for the duplicate

plots of Bonza 55 are considered to be the result of
random sampling errors of remote distances from the
point source where disease quantities were extremely
small." Fig. 3 of MacKenzie's paper shows that the
steeper gradient of Bonza 55 is based on 2 observations
only, and the flatter one on 4 observations. Thus, the
steeper gradient of Bonza 55 should be excluded from
consideration. The flatter one is the flattest of all
the gradients presented.

D Conclusions

The same result as derived here, a steeper gradient
with an increasing infection rate, was obtained by Van
den Bosch et al. (1988a). All models which describe
focal disease development in time and space are
consistent in this result. The opposite result of
Vanderplank is due to extrapolation of the `point'
model beyond its `domain of applicability'.

The interesting result of the experimental work by
MacKenzie (1976) is the observation, that below 50 % of
infected host tissue the flattening of the secondary
gradient, predicted by Gregory (1968), was not
observed, the gradient being defined as the first
derivative with space of the lesion density function at
a fixed position. Only when the disease severity on the
inoculated side of the measurement point reaches the
saturation level, gradients will flatten. MacKenzie
expressed the opinion that the flattening of the
gradient can occur at high disease levels. His opinion
is consistent with the results of our numerical
analysis. If on the other hand we define the gradient
as the first derivative measured at the 50 % level of
disease severity, then the gradient is constant during
each of the runs.

4.2.2 EPIMUL 76 - two spatial dimensions and time

One of the first models simulating focus formation in two-dimensional space and time was EPIMUL (Kampmeijer and Zadoks, 1977). Its theoretical basis is formed by the following assumptions:

1. space is compartmentalized into 400 (20 x 20) compartments,
2. disease develops uniformly in each compartment,
3. spores are produced by lesions with constant rate (DMFR) during the period from p till $p+i$ after lesion initialization,
4. after liberation, spores are distributed over space during one day by turbulent diffusion, and then suddenly deposited,
5. $(1 - x_t)$ of the deposited spores successfully infect the host, where x_t is the diseased fraction of host plant area.

The model was programmed in FORTRAN. A series of simulation runs with different sets of parameters gave several phytopathologically important results, such as:

1. gradients of the disease severity in dependence of the spore distribution parameter HALRIB (= HALF/RIBB, where HALF is the distance between the spore source and the place where the density is half the density at the source and RIBB is a side of a compartment),
2. displacement velocity of the focal front in dependence of a daily multiplication factor (DMFR, number of offspring produced per sporulating lesion per day) and of the spore distribution parameter HALRIB,
3. pictures of the diseased region in dependence on time and on the spore distribution parameter HALRIB.

A Parameters

Some of the parameters needed by the 'diffusion model' can be directly taken from EPIMUL, the others must be translated.

1. Unchanged parameters.
    - p - latency period   [T]
    - i - infectious period   [T]
2. Translated parameters.
    - L_{max} - the maximum number of lesions per compartment $[NL^{-2}]$
    - R - number of spores produced by a single sporulating lesion per unit of time $[NN^{-1}T^{-1}]$
    - E - infection efficiency   [1]
    - D - diffusion coefficient $[L^2T^{-1}]$
    - δ - rate of spore deposition $[T^{-1}]$

Ad L_{max}.

$$L_{max} = \frac{LAI}{AREA} \; (\text{area of compartment}) \qquad (4.5)$$

where LAI is the leaf area index and AREA is the area of a single lesion.

Ad R, E. The number of daughter lesions produced per mother lesion per day, DMFR in EPIMUL, is equal to the product of two parameters of the 'diffusion theory':

$$DMFR = R \; E \qquad (4.6)$$

Because no spores are removed from the epidemic in EPIMUL and all plants are totally susceptible, E = 1.

Ad D. The 'distribution' parameter of EPIMUL - HALF should be translated into terms of D. During one simulation day, spores are distributed according to a Gauss function with variance

$$\sigma^2 = 2\,D$$

then,

$$D = \frac{\sigma^2}{2} \qquad (4.7)$$

The 'dispersion' parameter of EPIMUL is HALF - the distance in meters between the spore source and the place where the density is half of the density at the source. The following relation between σ and HALF can be written (rearanged equation (2.7) of Kampmeijer and Zadoks, 1977):

$$\sigma = \frac{\text{HALF}}{\sqrt{2 \cdot (\ln 2)}} \qquad (4.8)$$

Substituting (4.8) into (4.7) leads to the following equation on D:

$$D = \frac{\text{HALF}^2}{4 \cdot (\ln 2)} \qquad (4.9)$$

Ad δ. There is no assumption in EPIMUL, which allows to assess the value of the rate of spore deposition, but the obvious choice is to use δ in the order of magnitude of 1 [day^{-1}] (almost all spores are deposited within 1 day).

B Results

To compare results of the two models, twelve runs of the computer programme PODESS (Appendix A) were performed on a VAX 8600 computer. For all runs, the solution field (100 m x 100 m) was discretized into 21 x 21 grid points. The initial infection by one spore occured at the centre of the field. The following parameters had the same values in all runs:

1. p = 8 [days],
2. i = 8 [days],
3. E = 1,
5. δ = 2 [1/day],
6. LAI   = 5,
7. AREA = 10 $[\text{mm}^2]$,
8. R varies with runs from 2. to 50. [spores per sporulating lesion per day].

The results of these runs were compared to the results of EPIMUL, Tables 2 and 3 and Fig. 13 of Kampmeijer and Zadoks (1977). The results of EPIMUL depend on the value of the parameter HALRIB. Substituting the value of HALF (calculated from HALRIB), D can be calculated from (4.9). The most interesting results of EPIMUL were obtained for values of HALRIB = 0.2, 0.4, 0.5, 1.0, 2.0, and 10.0. Applying

Table 4.5. A comparison of the 'diffusion model' and EPIMUL. Gradient values, at severity level 0.05, obtained by EPIMUL and by the 'diffusion model' for corresponding values of HALRIB (EPIMUL) and D ('diffusion model'). Parameter values: R = 10, others as in text.

EPIMUL		'Diffusion model'	
HALRIB	gradient	D	gradient
0.2	- 0.45	0.36	- 0.86
0.5	- 0.19	2.25	- 0.11
1.0	- 0.065	9.0	- 0.057
2.0	- 0.005	36.1	- 0.051

equation (4.9) the following values of the diffusion coefficient were calculated: D = 0.36, 1.45, 2.25, 9.0, 36.1, and 902.

Table 4.5 compares the results of Table 2 in Kampmeijer and Zadoks with the results of the 'diffusion model'. The two models allow to calculate the displacement velocities of the disease front (at constant severity level), Table 4.6. The plots of disease intensity (Fig. 4.4) made according to the 'diffusion' theory were compared to those of EPIMUL (Fig. 13 of Kampmeijer and Zadoks, 1977). These figures show the development of five focal epidemics for various values of HALRIB. The two sets of figures look very similar, thus indicating qualitative consistency of the two models.

Table 4.6. A comparison of the 'diffusion model' and EPIMUL. The velocity of frontal displacement, in compartments per day. Values of parameters other than R and D are given in the text.

EPIMUL			'Diffusion model'		
DMFR	HALRIB	velocity	R	D	velocity
2.0	0.2	0.03	2.0	0.36	0.07
	0.4	0.10		1.45	0.10
	1.0	0.22		9.0	0.21
10.0	0.2	0.08	10.0	0.36	0.10
	0.4	0.14		1.45	0.15
	1.0	0.28		9.0	0.27
50.0	0.2	0.09	50.0	0.36	0.14
	0.4	0.17		1.45	0.19
	1.0	0.33		9.0	0.33

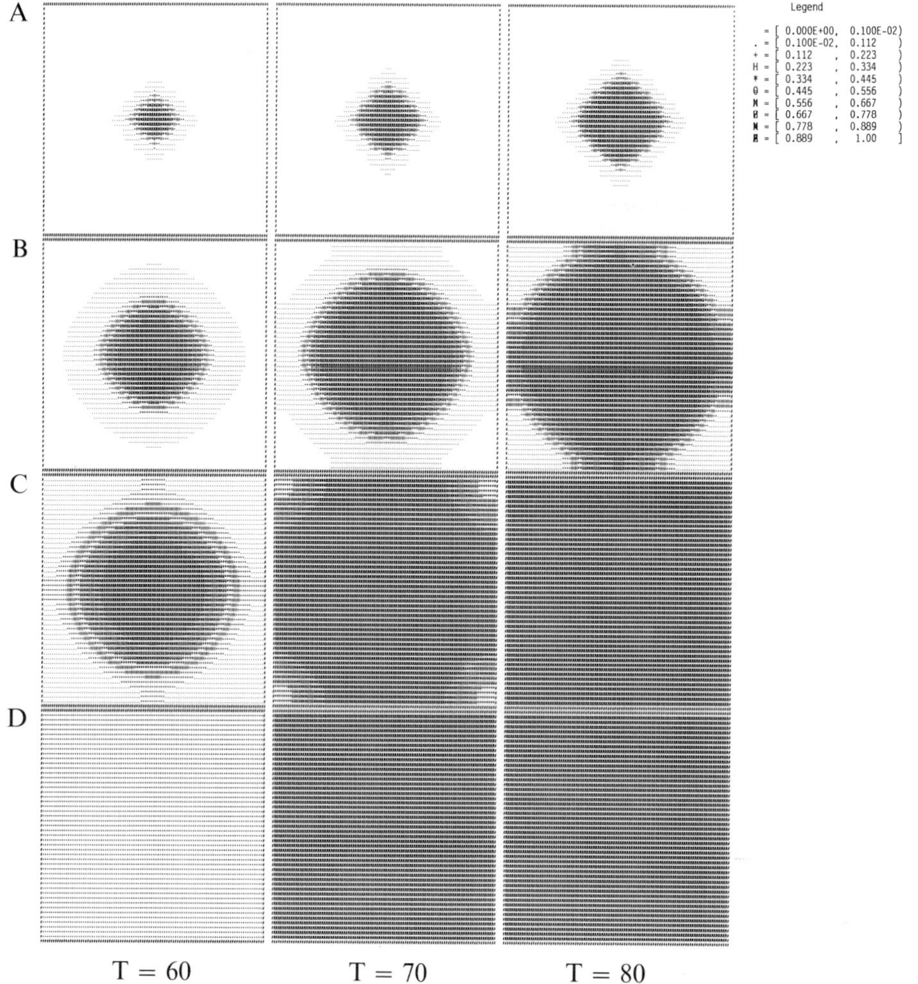

Fig 4.4. Development of simulated focal disease for time T = 60, 70 and 80. X- and Y-axes are distances from 0 to 100 m, intensity of printed points reflects the fraction of the host surface covered by lesions. The 'diamond' shapes of the diseased area are due to discretization of space by the numerical method of solution and by the method of plotting. Values of the diffusion coefficient are: **A**, $D = 0.36$; **B**, $D = 9$; **C**, $D = 36.1$; **D**, $D = 902$. In all cases $R = 10$, other parameter values as in text.

C Discussion

The two models are essentially different in their assumptions about the distribution of spores, though both are subsumed under the Diekmann-Thieme theory. Therefore, qualitative rather than quantitative consistency of the two models should be expected. Comparison of the results obtained by numerical analysis of the equations of the 'diffusion theory' and those presented by Kampmeijer and Zadoks (1977) confirms this opinion. The fundamental difference between the two models can be explained as follows. The 'diffusion model' takes into account two processes that lead to a decrease of the density of air-borne spores with the distance from the plant of origin: (1) turbulent diffusion and (2) deposition with constant probability per unit of time. Diffusion and continuous deposition together lead to the Bessel form of the spore deposition density (Broadbent and Kendall, 1953; Van den Bosch et al., 1988a, b). EPIMUL assumes that all spores stay in the air for a fixed time during diffusion and that after that time they are suddenly deposited.

The influence of different lesion distribution functions becomes evident in the results of Table 4.5. The Bessel function is more 'peaked' than the Gauss function, which is equivalent to a steeper gradient near the point of origin, and a flatter one in the distal region. This results in a steeper gradient for the 'diffusion model' than that of EPIMUL in the first row of Table 4.5, and a flatter one in the second and the third row. The phenomenon can be explained as follows. Severity level 0.05, at which the gradient values were calculated, is rather low, so that, in the cases compared at the second and the third lines of Table 4.5, the calculations were performed in the region of a flat Bessel function gradient. In the case considered on the first line of this table, the severity function was so `peaked', that the region of gradient assesment contained the steeper part of the Bessel function. The steeper gradient in the fourth row of Table 4.5 has a similar reason.

Good qualitative consistency of the two models was shown by the displacement velocity of the focal front (Table 4.6). However, some differences can be observed for low values of HALRIB (for EPIMUL) and corresponding values of δ (for the `diffusion model'). These differences are due to the different spore distribution functions of the two models. It can be shown, using the perturbation expansions described by Van den Bosch et al. (in prep.), that for low velocities, the velocity of focus expansion depends only on the variance of the spore distribution function, and not on its shape. The steepness of the focal front is far more sensitive to the shape of the distribution function than its displacement velocity.

EPIMUL and the `diffusion model' show good qualitative and fair quantitative consistency. Differences in the assumptions on spore dispersal are responsible of the observed minor discrepancies. On *a priori* grounds we may state that the spore dispersal mechanism of the `diffusion model' reflects reality better than that of EPIMUL. Van den Bosch et al. (1988b, c) review experimental material confirming this view.

4.3 COMPARISON WITH EXPERIMENTAL RESULTS

4.3.1 Experimental results of downy mildew on spinach

Comparison of the `diffusion model' to the models known from the literature may be good, experimental validation is better. Numerical results of the `diffusion model' were compared to experimental data for downy mildew (*Peronospora farinosa*) on spinach (*Spinacia oleracea*), and to analytical results obtained by applying the Diekmann-Thieme theory (Van den Bosch et al., 1988c). On the basis of field experiments with spinach (cv. Noorman) performed in 1983 on 1.5 m x 1.5 m plots inoculated at the centres and a similar experiment performed in the greenhouse in 1984 with cv. Huro, Van den Bosch et al. (1988c) assessed the necessary parameter values and the velocity of focus expansion. Using the Diekmann-Thieme theory together with experimentally determined parameter values, they calculated the expected velocity of focus expansion. The difference between the two velocity values, observed and predicted, is within the experimental error. Similarly, the velocity of focus expansion was calculated by the `diffusion model' using the parameter values given by Van den Bosch et al. (1988c). The calculated velocity was compared to the observed one, and to the velocity calculated by numerical solution of the Diekmann-Thieme speed equation.

A Parameters

Van den Bosch et al. (1988c, and personal communication), used the following parameter values which here are used as input data for the numerical solution of the equations of the `diffusion theory':

1. p - latency period = 7 days   [T]
2. $R(t-p)$ E - number of daughter lesions produced per sporulating mother lesion = 0.041, 1.44, 0.33, 0.65, 0.20, 0.18, 0.13, 0.15, 0.055 (spores per sporulating lesion per day) for the 1st till 9th day of sporulation, respectively $[NN^{-1}T^{-1}]$

3. LAI   - leaf area index = 5   [1]

4. AREA   - area of a single lesion = 1 cm^2 $[L^2]$

5. D - diffusion coefficient $[L^2 T^{-1}]$

6. δ - rate of spore deposition $[T^{-1}]$.

The last two parameters, D and δ, cannot be estimated separately from available data. One dispersion parameter, the width, ρ, of the Bessel contact distribution was measured by Van den Bosch et al. (1988c). According to Williams (1961) the parameter ρ^2 as measured by the mean square value of the distance from the source of spores equals:

$$\rho^2 = \frac{4\,D}{\delta} \tag{4.13}$$

where D is the diffusion coefficient and δ is the deposition rate of spores. With ρ = 0.163 ± 0.047 m (see Table 1 of Van den Bosch et al., 1988c), this leads to

$$D\,/\,\delta = 0.0066 \ [m^2]$$

Taking into account the standard deviation of the estimation of ρ, this ratio varies from 0.0034 m^2 to 0.011 m^2.

Because only the ratio D/δ influences the solution, arbitrary values of D and δ, which keep this ratio constant, can be chosen. The values δ = 2 [1/day] and D = 0.013 [m^2/day] were used in the run of the 'diffusion model' discussed above. This choice leads to an adequate value of D/δ while the equations do not yet become stiff, so that their numerical solution is relatively fast.

B Results

The numerical solution of the equations of the

`diffusion theory' was performed on a VAX 8600 computer, using the computer programme PODESS (Appendix A).

The `solution' region was a field of 1.5 m x 1.5 m, `inoculated' at the centre with 100 spores. During 100 `solution days' the results were printed and plotted every tenth day. They allow calculating the velocity of displacement of the focal front. The calculated velocity was 0.024 m/day (for D = 0.013 [m^2/day]), For ρ = 0.163 ± 0.047 m, the confidence limits are 0.014 m/day (for D = 0.0068 [m^2/day]) and 0.041 m/day (for D = 0.022 [m^2/day]). The velocity is close to the result obtained by Van den Bosch et al. (1988c) calculated on the basis of the Diekmann-Thieme theory, 0.03 ± 0.024 m/day, and - more important - with the experimental result, 0.023 ± 0.002 m/day. The discrepancy between the result of the `diffusion model' and the result obtained by Van den Bosch et al. is due to the discretization error inherent in the fit of the spore production kernel used by the `diffusion model'.

4.3.2 Mixtures of susceptible and resistant varieties

Vulnerability of crops can be decreased in a variety of ways. One is mixing resistant and susceptible varieties. The effectiveness of such mixtures was studied experimentally (e.g. Zadoks, 1958; Mundt et al., 1986a) and by computer simulation (Kampmeijer and Zadoks, 1977; Mundt et al., 1986b, c). The rate of disease progress can be measured by the velocity of focus expansion, c_o. The dependence of c_o on the fraction of susceptibles in a mixture was determined experimentally (Buiel et al., in prep.) and analytically (Van den Bosch, personal comm.). Results of the two approaches were consistent; the velocity of focus expansion increases linearly with the logarithm of the percentage of suscepts in a mixture.

The experiment (Buiel et al., in prep) was performed in 1987 with twelve plots of 3 m x 3 m. Four combinations of mixtures of susceptible (cv. Okapi) and resistant (cv. Sarno) wheat were planted in three replications. The proportions of susceptible to resistant plants in the mixtures were: 1:0, 1:1, 1:2, and 1:4 (percentages of suscepts: 100%, 50%, 33%, and 20%). Each plot contained 11 x 11 wheat hassocks. Plots were inoculated in the center by planting 2 additional clumps of the susceptible cultivar inoculated by stripe rust (*Puccinia striiformis*); after a few days these clumps were removed. The velocity of focus expansion, c_o, was assessed for each plot on the basis of lesion counts per hassock. The relation between c_o and the logarithm of the percentage of suscepts was determined.

A Parameters

Some parameters required by the `diffusion' theory were measured during the experiment. The values of the others were guesstimated.

Directly measured were latency period, p = 17 days, and infectious period, i = 21 days. Using the contact distribution, assessed from the distribution of the first generation lesions, the parameter of the Bessel distribution i.e. the mean square value of the distance, ρ^2, was estimated by the method of Williams (1961). Then, using equation (4.16), the ratio D/δ was calculated. The third column of Table 4.7 gives the values of ρ for each plot. Assuming a constant value of the diffusion coefficient D = 0.015 [m^2/ day], the value of the rate of deposition, δ, was estimated. The values of D and δ were chosen to maintain an appropriate value of D/δ and to use a relatively low value of δ (the value of δ influences computing time). The estimated values of δ are shown in column 4 of Table 4.7. As the reproductivity parameters were not measured, the value of 10 daughter lesions per mother lesion (Zadoks, 1961) was assumed for plots with 100% suscepts. This value gives approximately $R = 0.5$ daughter lesions per mother lesion per day. Infection efficiency E was then assumed to be equal to the fraction of suscepts in a mixture. Therefore, $E = 1.$, 0.5, 0.33 and 0.2 for 100%, 50%, 33% and 20% of suscepts in a mixture, respectively.

The constant parameters for all the runs were:
1. p = 17 [days],
2. i = 21 [days],
3. D = 0.015 [m^2/day],

Table 4.7. Experiment on focus expansion in cultivar mixtures of wheat (Buiel et al., in prep.). The percentage of suscepts in a mixture, the parameter of the Bessel distribution, ρ, and the rate of spore deposition, δ, for the twelve plots of the experiment.

Plot nr.	% of suscepts	ρ	δ
1	100	68	4.4
7	100	80	3.8
11	100	66	4.5
2	50	52	5.8
5	50	66	4.5
10	50	56	5.4
3	33	48	6.3
6	33	40	7.5
9	33	56	5.4
4	20	30	10.0
8	20	38	7.9
12	20	48	6.3

4. R =   0.5   [daughter lesions per mother lesion
                 per day],
5. AREA =   1   [mm^2] - area of a single lesion,
6. LAI =   4 - leaf area index.

For the first twelve runs, the values of δ are given
in Table 4.7, and the infection efficiency is equal to
the fraction of suscepts in a mixture. For three
additional runs δ = 4.4 (as for the first run) and E =
0.5, 0.33 and 0.2, respectively.

B Results

The `diffusion model' was run fifteen times. Twelve
runs were made with the values of δ derived from the
experimental determination of the parameter p of the
contact distribution, while D was assumed fixed. Three
additional runs were performed to study the influence
of varying the fraction of suscepts in a mixture, while
δ was kept fixed. The results produced by these runs
allowed to calculate the velocity of focus expansion,
c_0. Mean values for replications with the same
fractions of suscepts were calculated for the first
twelve runs. Then, the ratio of this mean velocity to
c_0 for plots with susceptibles only was calculated
(column 2 of Table 4.8). For the experimental results,
see column 1 of Table 4.8. Results of additional runs
together with the result of run 1 are shown in column 3
of Table 4.8.

Values in column 4 of Table 4.8 are lower than those
in column 3, a result which disagrees with that of
Section 4.2.1: The velocity of focus expansion should
be proportional to the logarithm of the mean free path
for absorption. The discrepancy seems to result from
the low level of disease severities at which the
calculations were performed, leading to values
calculated for the initial phase of focus formation,
when the velocities are not yet stable.

Table 4.8. Relative velocities of focus expansion for four proportions of susceptibles in a mixture of susceptible and resistant wheat plots. Results were calculated from experimental data (Buiel et al., in prep.), for twelve simulation runs with estimated values of δ (runs 1 to 12), and from four runs (1, 13, 14 and 15) with constant value of δ. Data in columns 2 and 3 are means of 3 replications.

Percentage of susceptibles	Experimental results	Calculated results, δ varies with runs	Calculated results, δ = 4.4
100	1.00	1.00	1.00
50	0.66	0.70	0.62
33	0.48	0.45	0.37
20	0.25	0.29	0.20

C Discussion

The 'diffusion theory' and the Diekmann-Thieme model are not fundamentally different. The Diekmann-Thieme model encomposes a family of models of which the 'diffusion theory' is just one member. Both models produce results consistent with experimental data.

The velocity of focus expansion gives information about the effectiveness of mixtures of resistant and susceptible varieties. Therefore, the Diekmann-Thieme model or the 'diffusion theory' can be used to predict the performance of mixtures. Application of the Diekmann-Thieme model (which is much simpler numerically) gives the best approximation of the velocity of focus expansion. The same result can be obtained by the `diffusion theory', but it needs considerable amount of computer time. Therefore, the above calculations were done only for the purpose of validation. The real advantage of the `diffusion theory' is in its ability to deal with transients, non uniform crop distribution, stochasticity, and so on. Examples will be given in Chapter 9.

4.4 DISCUSSION

The numerical analysis of the system of partial differential equations, which constitute the base of the `diffusion theory', showed qualitative consistency of theory and experimental data. Quantitative consistency with EPIMUL was fair. As special cases of the Diekmann-Thieme theory (a possible translation was presented in Chapter 3), the `diffusion theory' and the model of Van den Bosch et al. (1988a, b, c) are mutually consistent. The `diffusion theory' more accurately handles the processes governing spore distribution than Minogue and Fry's model or EPIMUL. Therefore, it is closer to reality than earlier models, as confirmed by good quantitative consistency with experimental data.

Chapter 4 discusses validation using existing information. The conclusion is that the `diffusion model' - and therewith the `diffusion theory' - is valid, that is `sound, defensible, well-grounded' (Concise Oxford Dictionary). Verification, i.e. providing proof that the `diffusion model' gives a `true' picture of reality, was not the objective of the present study.

The foregoing analysis shows the usefulness of the `diffusion theory' of focus development. To improve consistency between theory and experimental results, careful measurements are needed of all parameters required by the theory. The great number of high precision measurements needed seems to be a disadvantage of the theory. The problem can be solved by determining which parameters are most important to the `diffusion theory'. Treating only the measurements of these parameters with special care and using approximate values for the other parameters may decrease the effort needed to apply the `diffusion theory'. An attempt in this direction will be made in Chapter 5.

5 `SENSITIVITY' ANALYSIS BY MEANS
OF A UNIFORM ROTATABLE CENTRAL COMPOSITE DESIGN

5.1 INTRODUCTION

The phenomena which constitute the real world can often be formulated in the language of mathematics, i.e. in terms of a model consisting of a set of equations. The power of this approach is in its generality. Apart from independent variables, which change their value regularly, the equations contain parameters, which take certain values in particular cases. The mathematical model is used to `simulate' the behaviour of a system. A particular case is simulated by solving the equations with appropriate values of the parameters. From the simulation output we extract one or more numbers, which can be compared to experimental results. These numbers can be considered as the response of the model for the parameter values under consideration. The effect of a parameter on a result of a simulation run can vary from one parameter to another. The effect of any parameter does not only depend on the value of that parameter itself, but also on the values of other parameters. If, for given ranges of parameter values, the response cannot be decomposed into additive contributions of the separate parameters, the parameters are said to interact.

To assess the effect of a parameter on a response and to compare the relative effects of different parameters, a method is used called `sensitivity analysis'. Sensitivity analysis by means of varying individual parameters has been applied frequently in simulation studies (Zadoks, 1971; Rabbinge, 1976; de Wit and Goudriaan, 1978). A modeler varies a single parameter's value a little up and down (for example 10%) keeping other parameters constant and observes the

changes in a response relative to the variations in the
test parameter. When applied to a completely
deterministic model, `sensitivity analysis' helps to
judge the relative importance of a single measured
parameter in determining the response under the *ceteris
paribus* hypothesis. This kind of analysis can be also
done for the `diffusion theory', but because it
disregards interactions between parameters, another
method is proposed.

For the application of simulation methods to
agricultural systems, we have to consider carefully to
what extent the assumption of complete determination is
applicable. Basically, there are three types of
indeterminacy, (1) measurement noise in the response,
(2) stochasticity inherent in the process itself, and
(3) uncertainty in the model parameters. Measurement
noise is not considered here. Process stochasticity
takes two forms: (a) stochasticity due to a limited
number of individuals, and (b) variability of physical
circumstances (parameters) over time and space. The
first type of stochasticity is considered in Sections
8.5 and 9.3. The second type is addressed in Section
8.2 and throughout Chapter 9. Here, we deal with the
effect of changes of fixed parameters. The approach is
analogous to one followed by statisticians but it is
elaborated in a fully deterministic context.

The formal methods of sensitivity analysis as
developed by engineers do not help much. Usually, we
are interested in aspects of the response which do not
bear a simple relation to the first variation of the
solution of our equation with respect to the parameter
chosen. Morever, we want robust estimates over fairly
large ranges of the parameters, instead of purely local
results. Therefore, we fit a quadratic response surface
to the observed relation between an output quantity and
the parameter values under consideration. As a first
step we will carefully plan a number of simulation runs

which is as small as possible and yet allows to proceed
to the next step. Then we will fit a nonlinear function
to the results, and finally we will determine the
'importance' of each coefficient of the fitted
function.

5.2 THE METHOD

The method presented here assesses the coefficients
of a nonlinear function relating the variation of a
response to the single, squared and combined effects of
variations in parameters. The scaled uniform rotatable
central composite design (Box and Hunter, 1957;
Petersen, 1985) will be used to derive a series of
necessary simulation runs.

A linear function of parameters does not account for
the dependence of responses on possible interactions
between parameters. Therefore, a nonlinear function
must be used. The simplest one is the second order
function:

$$y = b_0 + \sum_{i=1}^{N} b_i \cdot x_i + \sum_{i=1}^{N} \sum_{\substack{j=1 \\ j \leq i}}^{N} b_{ij} \cdot x_i \cdot x_j \qquad (5.1)$$

where b_0, b_i and b_{ij} are coefficients, x_k ($k = i$ or $k = j$) is a kth independent variable (parameter), N is the
number of independent variables (parameters), and y is
the dependent or response variable. By means of a
multiple regression procedure (Draper and Smith, 1966;
Mosteller and Tukey, 1977; Jennrich, 1977), the
coefficients b_i and b_{ij} can be determined. Taking into
account their biological meaning and the influence on
the model's numerical response, the 'importance' of the
terms appearing in the expression (5.1) can be
determined. The model parameters (independent
variables) present in the 'important' terms are treated

as those with proven effect on the model. Special attention must be devoted to their estimation from field data.

5.3 THE EXPERIMENTAL DESIGN

How well we can choose the coefficients of the function (5.1) depends on the range of variation of parameter values, treated here as values of independent variables, and on the design of the set of simulation runs chosen. The qualification 'well', defined relative to the quality of the ensuing predictions, can be measured by the sum of squared deviations of predictions from measured responses. An appropriate design decreases the influence of the prediction errors of the parameter contributions on the quality of the values of the coefficients. A variety of experimental designs can be found in the literature (Cochran and Cox, 1957; Manczak, 1976; McLean and Anderson, 1984; Petersen, 1985).

As function (5.1) contains quadratic terms, $b_{ii} \cdot x_i^2$ (for $i = j$), the multilevel design is to be used (Box and Hunter, 1957; Manczak, 1976, Petersen, 1985). The uniform rotatable central composite design will be used here. It consists of:

1. a two-level factorial design which can be performed in one of two possible versions, (a) a full design (called 2^N, N being the number of parameters) with 2^N experiments, or (b) a fractional design (called 2^{N-M}, where N is the number of parameters and M takes some value $< N$) with 2^{N-M} experiments, in combination with

2. $2 \cdot N$ experiments at the 'axial points', and

3. N_0 experiments at the 'central point'

The terms 'axial point' and 'central point' will be explained in Section 5.3.1. Therefore, the complete uniform rotatable central composite design consists either of

$$L = 2^N + 2 \cdot N + N_0 \qquad\qquad (5.2)$$

simulation runs (experiments) for a full two-level design as the basic central composite design, or of

$$L = 2^{N-M} + 2 \cdot N + N_0 \qquad\qquad (5.3)$$

simulation runs (experiments) for a fractional two-level design as the basic central composite design.

5.3.1 Theory

In this section a short description of the uniform rotatable central composite design is given.

The 'behaviour' of a simulation model in the vicinity of a point in the N-dimensional parameter space, $\mathcal{P} = [x_1^o, \ldots, x_N^o]$, is to be examined. This point is the 'central point' of the experiment. Its neighbourhood (the region of the parameter space), where the response of the model must be examined, is an N-dimensional hyper-cuboid $[x_i^o - \Delta x_i, x_i^o + \Delta x_i]$, where x_i^o is the ith coordinate of $\mathcal{P}$ and Δx_i is a change of x_i. The particular value of Δx_i depends on the modeler's choice; the section $[x_i^o - \Delta x_i, x_i^o + \Delta x_i]$ should cover the range of values of the ith parameter which are interesting from a scientific point of view. Therefore, x_i^o and Δx_i take values which are determined by their biological context.

Normalization of variables

$$x_i \rightarrow z_i = \frac{x_i - x_i^o}{\Delta x_i}, \qquad i = 1, \ldots, N \qquad\qquad (5.4)$$

simplifies the notation, because z_i varies from -1 to

+1 and equals 0 for $x_i = x_i^o$. Points $z_i = \pm 1$ lay on the surface of a hypercube in the parameter space. Function (5.1) becomes:

$$y = c_0 + \sum_{i=1}^{N} c_i \cdot z_i + \sum_{i=1}^{N} \sum_{\substack{j=1 \\ j \leq i}}^{N} c_{ij} \cdot z_i \cdot z_j \qquad (5.5)$$

where c_0, c_i and c_{ij} are new coefficients.
   Function (5.5) contains

$$\mathfrak{N} = 1 + N + \frac{N \cdot (N + 1)}{2} \qquad (5.6)$$

coefficients, which are to be determined. This requires at least $\mathfrak{N}$ simulation runs. A full two-level design consists of 2^N experiments for all combinations of $z_i = \pm 1$ ($i = 1, \ldots, N$). A complete set of simulations for all combinations requires 2^N simulation runs (for instance for $N = 10$ we should make 1024 runs), though only $\mathfrak{N}$ are needed. Fortunately, this design can be reduced to a fractional two-level design with 2^{N-M} runs, for some value of M (Cochran and Cox, 1957; Cox, 1958; Finney, 1960). For $M = 1$ only one half (2^{N-1}) of the number of runs required by a complete design is to be made, for $M = 2$ one quarter (2^{N-2}), and so on. Plans for these and other designs can be found in Cochran and Cox (1957). Two-level designs allow to estimate the coefficients in linear (containing z_i) and mixed nonlinear (containing $z_i \cdot z_j$ for $i \neq j$) terms of a fitted function at the points of $\mathfrak{P}$ included in the design. All quadratic terms have values z_i^2 = +1, so that they are linearly dependent on a 'virtual variable' $z_0 \equiv 1$, which is introduced to calculate c_0 in (5.5)).
   Determination of the coefficients of the quadratic terms needs more points than only $z_i = \pm 1$, because two

points allow for the determination of straight line coefficients only. A quadratic term describes a curvilinear `behaviour' of a function. Therefore, determination of its coefficients needs at least three data points. A good choice is to use the central composite design with five points, because it allows to fulfil some additional requirements.

A special case of the central composite design is the uniform-rotatable design (Box and Hunter, 1957; Manczak, 1976; Petersen, 1985). The latter design ensures equal mean square errors of the response estimated by the fitted function in every direction on an N-dimensional sphere with center at $z_i = 0$ ($i = 1, \ldots, N$). This means that the estimated response is a function only of the distance of a point from the center of the design. It also ensures that for the central point and the sphere with radius 1, the mean square errors of the values estimated by the function (5.5) are equal. Therefore, the mean square error is almost constant for all spheres with radii between 0 and 1. This implies an approximately uniform precision over the parameter space spanned by radii $\rho = 0$ to $\rho = 1$. The uniform rotatable central composite design requires, in addition to the requirements of a two-level design, $2 \cdot N$ simulation runs at the so called `axial point' for every single parameter, whereas other parameters are at their centers ($z_j = 0$ for $j \neq i$), and $N_0$ runs at the center ($z_i = 0$ for $i = 1, \ldots, N$). The `axial point' is a point in parameter space, of which the distance from the center ($z_i = 0$, $i = 1, \ldots, N$) allows to fulfil the condition of rotatability. The number of runs at the central point, N_0, is needed to fulfil the condition of uniform precision. Therefore, a uniform-rotatable design consists of L runs, where L is determined by equation (5.2) or (5.3). If, as in the present situation, parameters (independent variables) are determined without random variation, all N_0 runs at the central point will give exactly the same result. Therefore, these runs can be replaced by one run at the central point, of which the result will be used in the analysis with the weight N_0.

It can be proven (Box and Hunter, 1957) that, for a rotatable (not necessarily uniform) design, the `axial point' is the point ($z_i = \alpha$; $z_j = 0$ for all $j \neq i$) with

$$\alpha = \sqrt[4]{2^N} \tag{5.7}$$

for a full two-level design as the basic central composite design, and

$$\alpha = \sqrt[4]{2^{N-M}} \tag{5.8}$$

for a fractional two-level design as the basic central composite design.

A long derivation (Box and Hunter, 1957) leads to the equation for N_0 (the number of runs at the central point or the weight of the result of a single run at the central point):

$$N_0 = \mu \, (2^{N-M} + 2^{(N-M)/2+2} + 4) - 2N - 2^{N-M} \tag{5.9}$$

where μ is a coefficient, depending on N, to be calculated with the assumption that the mean square errors of the values estimated by function (5.5) in the center of the design and on the sphere with radius 1 around the center are all equal. This condition characterizes uniform designs. The value of the coefficient μ is slightly below 1. It is tabulated by Box and Hunter (1957, Table 1) for values of N from 2 to 8.

5.4 THE DESIGN FOR THE `DIFFUSION THEORY'

The results of simulation runs applying the
`diffusion theory' depend on six parameters. These are:
1. D - diffusion coefficient $\qquad$ $[L^2 T^{-1}]$
2. δ - deposition rate $\qquad$ $[T^{-1}]$
3. R - number of spores produced by a
   sporulating lesion per unit of time $[T^{-1}]$
4. p - latency period $\qquad$ $[T]$
5. i - infectious period $\qquad$ $[T]$
6. E - infection efficiency $\qquad$ $[1]$
7. U - linear size of a square field $\qquad$ $[L]$.

These six parameters can be combined into three
dimensionless quantities. The following combinations
are made:
1. $\mathfrak{R} = R \cdot E \cdot i$ (number of daughter lesions produced
      per sporulating mother lesion),
2. $\mathfrak{I} = i \: / \: p$ (ratio of infectious to latency period),
3. $\mathfrak{U} = U \: / \: \sqrt{D/\delta}$ (ratio of field length to the width
      of the contact distribution).

The contact distribution (sensu Van den Bosch et al.,
1988a, b) measures the range of the spore dispersal.

Another dimensionless quantity must be chosen as the
response of the `diffusion model'. Two values measuring
the diseases spread are good candidates: (1) the scaled
velocity of focus expansion and (2) the total number of
lesions present in a field at a certain time.

5.4.1 The number of simulation runs

The number of coefficients of function (5.5) is
given by (5.6). For the `diffusion theory' $N = 3$ ($\mathfrak{R}$, $\mathfrak{I}$,
$\mathfrak{U}$), so that

$$\mathfrak{N} = 1 + 3 + \frac{12}{2} = 10. \qquad (5.10)$$

Determination of $\mathfrak{N}$ coefficients needs $L \geq 10$ simulation

5.5 THE RESULTS

Using empirical knowledge, the following ranges of the three dimensionless combinations of parameters of the 'diffusion theory' were chosen (for an explanation of the symbols used, see Section 5.4):

1. $\mathfrak{R}$: 3   -   27,
2. $\mathfrak{I}$: 0.6 -   2,
3. $\mathfrak{U}$: 10   - 1000.

The ranges of $\mathfrak{R}$ and $\mathfrak{I}$ are easy to interpret for epidemiologists. The range of $\mathfrak{U}$ was determined by choosing a field side length of 100 m under the assumption that dispersal distances from 0.1 to 10 m are to be considered for focus formation.

As the values of $\mathfrak{R}$ and $\mathfrak{U}$ grow exponentially rather than linearly, $\log_3 \mathfrak{R}$ and $\log_{10} \mathfrak{U}$ were used. Finally, the ranges become:

1. $\log_3 \mathfrak{R}$: 1   - 3,
2. $\mathfrak{I}$:   0.6 - 2,
3. $\log_{10} \mathfrak{U}$: 1   - 3.

Transformation (5.4) changed these ranges into the standard ranges from -1 to +1. According to (5.16), the 'axial points' were in -1.68 and +1.68. The values: -1.68, -1, 0, 1, and +1.68 were transformated, by application of (5.17), to:

1. $\mathfrak{R}$: 1.4, 3,   9,   27,   57,
2. $\mathfrak{I}$: 0.1, 0.6, 1.3,   2,   2.5,
3. $\mathfrak{U}$: 2.1, 10, 100, 1000, 4786.

These values were used to design 20 runs of the 'diffusion model' according to Table 5.1. Translation into the original parameters of the 'diffusion theory' gave the actual values used (Table 5.2). The space grid of 11 x 11 points was used.

The total number of lesions present in the field, $\mathfrak{L}$, at time T = 13, 25, and 50, was used as the response of the 'diffusion model'. These time values were chosen because they are 1.3, 2.5 and 5 times higher than the

96

5.4 THE DESIGN FOR THE `DIFFUSION THEORY'

The results of simulation runs applying the `diffusion theory' depend on six parameters. These are:
1. D - diffusion coefficient $[L^2 T^{-1}]$
2. δ - deposition rate $[T^{-1}]$
3. R - number of spores produced by a sporulating lesion per unit of time $[T^{-1}]$
4. p - latency period $[T]$
5. i - infectious period $[T]$
6. E - infection efficiency $[1]$
7. U - linear size of a square field $[L]$.

These six parameters can be combined into three dimensionless quantities. The following combinations are made:
1. $\mathfrak{R} = R \cdot E \cdot i$ (number of daughter lesions produced per sporulating mother lesion),
2. $\mathfrak{I} = i \, / \, p$ (ratio of infectious to latency period),
3. $\mathfrak{U} = U \, / \, \sqrt{D/\delta}$ (ratio of field length to the width of the contact distribution).

The contact distribution (sensu Van den Bosch et al., 1988a, b) measures the range of the spore dispersal.

Another dimensionless quantity must be chosen as the response of the `diffusion model'. Two values measuring the diseases spread are good candidates: (1) the scaled velocity of focus expansion and (2) the total number of lesions present in a field at a certain time.

5.4.1 The number of simulation runs

The number of coefficients of function (5.5) is given by (5.6). For the `diffusion theory' $N = 3$ ($\mathfrak{R}$, $\mathfrak{I}$, $\mathfrak{U}$), so that

$$\mathfrak{N} = 1 + 3 + \frac{12}{2} = 10. \qquad (5.10)$$

Determination of $\mathfrak{N}$ coefficients needs L $\geq$ 10 simulation runs.

5.4.2 The design

A uniform rotatable central composite design of an experiment (with a full two-level design as its basic part) consists of

$$L = 2^N + 2 \cdot N + N_o \qquad (5.11)$$

simulation runs, where N is the number of parameters, and N_o is the number of simulation runs which will be performed at the center ($z_i = 0$, $i = 1, \ldots, N$). N_o will be calculated in the present section. For $N = 3$, L from (5.11) becomes

$$L = 2^3 + 6 + N_o. \qquad (5.12)$$

Assessment of all the coefficients of function (5.5) needs at least 10 simulation runs (equation (5.10)). Therefore

$$L \geq 10 \qquad (5.13)$$

Substituting (5.12) in (5.13) we obtain

$$2^3 + N_o \geq 4 \qquad (5.14)$$

N_o is nonnegative, so that inequality (5.14) is satisfied for any arbitrary value of N_o.

The next step is the determination of N_o. Equation (5.9) with $N = 3$ and $M = 0$ gives

$$N_o = \mu \, (2^3 + 2^{7/2} + 4) - 6 - 2^3$$
$$= 23.31 \cdot \mu - 14 = 19.55 - 14 \approx 6 \qquad (5.15)$$

where μ for $N = 3$ equals 0.8385 (Box and Hunter, 1957).

The last unknown is the `axial point' value α, which can be calculated from (5.7):

$$\alpha = \sqrt[4]{2^3} \approx 1.68 \tag{5.16}$$

Introducing $N = 3$ and $N_o = 6$ into (5.11), the number of simulation runs L = 20 is obtained (Table 5.1).

For these simulation runs independent variables x_i are used, which are determined from the values z_i in Table 5.1 by the reverse transformation of equation (5.4):

$$x_i = x_i^o + z_i \cdot \Delta x_i \tag{5.17}$$

Table 5.1. Sensitivity analysis. The uniform-rotatable design for the `diffusion theory'.

no.	z_1	z_2	z_3
1	-1	-1	-1
2	-1	-1	+1
3	-1	+1	-1
4	-1	+1	+1
5	+1	-1	-1
6	+1	-1	+1
7	+1	+1	-1
8	+1	+1	+1
9	$-\alpha$	0	0
10	$+\alpha$	0	0
11	0	$-\alpha$	0
12	0	$+\alpha$	0
13	0	0	$-\alpha$
14	0	0	$+\alpha$
15 - 20	0	0	0

5.5 THE RESULTS

Using empirical knowledge, the following ranges of
the three dimensionless combinations of parameters of
the 'diffusion theory' were chosen (for an explanation
of the symbols used, see Section 5.4):

1. $\Re$: 3 - 27,
2. $\Im$: 0.6 - 2,
3. $\mathfrak{U}$: 10 - 1000.

The ranges of $\Re$ and $\Im$ are easy to interpret for
epidemiologists. The range of $\mathfrak{U}$ was determined by
choosing a field side length of 100 m under the
assumption that dispersal distances from 0.1 to 10 m
are to be considered for focus formation.

As the values of $\Re$ and $\mathfrak{U}$ grow exponentially rather
than linearly, $\log_3\Re$ and $\log_{10}\mathfrak{U}$ were used. Finally, the
ranges become:

1. $\log_3\Re$: 1 - 3,
2. $\Im$:   0.6 - 2,
3. $\log_{10}\mathfrak{U}$: 1 - 3.

Transformation (5.4) changed these ranges into the
standard ranges from -1 to +1. According to (5.16), the
'axial points' were in -1.68 and +1.68. The values:
-1.68, -1, 0, 1, and +1.68 were transformated, by
application of (5.17), to:

1. $\Re$: 1.4, 3, 9, 27, 57,
2. $\Im$: 0.1, 0.6, 1.3, 2, 2.5,
3. $\mathfrak{U}$: 2.1, 10, 100, 1000, 4786.

These values were used to design 20 runs of the
'diffusion model' according to Table 5.1. Translation
into the original parameters of the 'diffusion theory'
gave the actual values used (Table 5.2). The space grid
of 11 x 11 points was used.

The total number of lesions present in the field, $\mathfrak{L}$,
at time T = 13, 25, and 50, was used as the response of
the 'diffusion model'. These time values were chosen
because they are 1.3, 2.5 and 5 times higher than the

Table 5.2. Sensitivity analysis. The values of the 'diffusion theory' parameters used in 20 runs of the uniform-rotatable design for sensitivity analysis.

no.	E	R	i	p	D	δ	U
1	1	0.5	6	10	100	1	100
2	1	0.5	6	10	1	100	100
3	1	0.15	20	10	100	1	100
4	1	0.15	20	10	1	100	100
5	1	4.5	6	10	100	1	100
6	1	4.5	6	10	1	100	100
7	1	1.35	20	10	100	1	100
8	1	1.35	20	10	1	100	100
9	1	0.11	13	10	10	10	100
10	1	4.38	13	10	10	10	100
11	1	9	1	10	10	10	100
12	1	0.36	25	10	10	10	100
13	1	0.69	13	10	229	0.1	100
14	1	0.69	13	10	0.1	229	100
15-20	1	0.69	13	10	10	10	100

value of the latency period, p. The variable $\mathfrak{L}$ was used as the dependent variable and $\mathfrak{R}$, $\mathfrak{I}$, and $\mathfrak{U}$ were used as the independent variables in fitting of function (5.1). The values of these input variables and the resulting response values for 15 runs are shown in Table 5.3.

As the parameter values for runs 15 to 20 are identical, the results of the 15th run, that of the 'central point', were used $N_o = 6$ times in the least square fitting procedure. The runs were performed on a VAX 785 computer using the software package PODESS

Table 5.3. Sensitivity analysis. The values of the independent variables, $\mathfrak{R}$, $\mathfrak{S}$, and $\mathfrak{U}$, and three values of the dependent variable, $\mathfrak{L}(13)$, $\mathfrak{L}(25)$, and $\mathfrak{L}(50)$ at time T = 13, 25, and 50, respectively, for 20 runs of the uniform-rotatable design for sensitivity analysis of the 'diffusion theory'.

no.	$\mathfrak{R}$	$\mathfrak{S}$	$\mathfrak{U}$	$\mathfrak{L}(13)$	$\mathfrak{L}(25)$	$\mathfrak{L}(50)$
1	3	0.6	10	3	7	67
2	3	0.6	1000	3	9	85
3	3	2	10	1	4	18
4	3	2	1000	2	4	19
5	27	0.6	10	15	274	152955
6	27	0.6	1000	19	451	314471
7	27	2	10	5	44	5336
8	27	2	1000	6	61	8328
9	1.4	1.3	100	1	3	6
10	57	1.3	100	18	438	445895
11	9	0.1	100	10	91	42243
12	9	2.5	100	2	9	163
13	9	1.3	2.1	2	9	109
14	9	1.3	4786	4	20	777
15-20	9	1.3	100	4	19	744

(Appendix A). The second order function (5.1) was fitted to the results by the least square method using the GLM procedure from SAS.

Four functions of type (5.1) for the response at three times T = 13, 25, and 50 were tested: (1) $\mathfrak{L}(T)$ depending on $\mathfrak{R}$, $\mathfrak{S}$, and $\mathfrak{U}$, (2) $\log_{10}\mathfrak{L}(T)$ depending on $\mathfrak{R}$, $\mathfrak{S}$, and $\mathfrak{U}$, (3) $\mathfrak{L}(T)$ depending on $\log_{10}\mathfrak{R}$, $\mathfrak{S}$, and $\log_{10}\mathfrak{U}$, and (4) $\log_{10}\mathfrak{L}(T)$ dependence on $\log_{10}\mathfrak{R}$, $\mathfrak{I}$, and $\log_{10}\mathfrak{U}$. Coefficients of the approximating functions and the sums of squared residuals were compared. The best fit was obtained for the fourth combination. Therefore the following second order approximaton function was chosen:

$$\log_{10}\mathfrak{L}(T) = b_0 + b_1 \log_{10}\mathfrak{R} + b_2 \mathfrak{I} + b_3 \log_{10}\mathfrak{U} +$$
$$b_{12} (\log_{10}\mathfrak{R}) \mathfrak{I} + b_{13} (\log_{10}\mathfrak{R}) (\log_{10}\mathfrak{U}) +$$
$$b_{23} \mathfrak{I} (\log_{10}\mathfrak{U}) + b_{11} (\log_{10}\mathfrak{R})^2 +$$
$$b_{22} \mathfrak{I}^2 + b_{33} (\log_{10}\mathfrak{U})^2 \qquad (5.18)$$

where b_0, b_1, b_2, b_3, b_{12}, b_{13}, b_{23}, b_{11}, b_{22}, b_{33} are coefficients which values are given in Table 5.4.

The coefficients of all linear and quadratic terms differ non-negligably from zero, which indicates a great influence of $\mathfrak{R}$, $\mathfrak{I}$ and $\mathfrak{U}$ on the total number of lesions present in a field, $\mathfrak{L}$. The coefficients of two interaction terms, b_{13} and b_{23}, do not differ non-negligably from zero, but a third one, b_{12}, does. Sums of squared residuals are low, which means that the total number of lesions is 'well' described by the particular quadratic function chosen.

The sign of coefficient b_3 is positive what means that the total number of lesions present in the field grows with an increase of $\mathfrak{U}$. As $\mathfrak{U}$ is the ratio of field size to the width of the contact distribution, a higher value of $\mathfrak{U}$ is equivalent to a lower width of the contact distribution. The positive sign of b_3 indicates that some spores arriving at the field boundary are lost (blown outside the field). This is the result of the boundary condition chosen, see Section 3.5.4. A higher $\mathfrak{U}$ means that a higher proportion of spores stays within the field, and therefore initializes more lesions.

Table 5.4. Values of the coefficients of function (5.18) fittted to the results of simulation runs planned according to the uniform rotatable design for sensitivity analysis of the `diffusion theory'. Symbols are as in function (5.18), SSR stands for the `sum of squared residuals'. One asterisk means significance at $P \leq 0.05$, two asterisk mean significance at $P \leq 0.01$, if the prediction errors would have been due to independent Gaussian distributed measurement noise.

Coefficient	Values at time		
	T = 13	T = 25	T = 50
b_0	0.16	0.35*	0.66
b_1	0.34	0.94**	2.64**
b_2	-0.38**	-0.39**	-1.17**
b_3	0.25*	0.25**	0.65*
b_{12}	-0.16*	-0.37**	-0.68**
b_{13}	0.027	0.056	0.1
b_{23}	-0.0083	-0.027	-0.035
b_{11}	0.27**	0.45**	0.6**
b_{22}	0.11**	0.15**	0.39**
b_{33}	-0.049**	-0.048**	-0.14**
SSR	0.032	0.024	0.22

The influence of $\mathfrak{R}$ on $\mathfrak{L}$ is easy to understand, because $\mathfrak{R}$ is the total number of offspring produced by a lesion. $\mathfrak{I}$ is inversly proportional to $\mathfrak{L}$, as shown by negative values of b_2. This means that a shorter infectious period, when the total number of daughter lesions per mother lesion is constant, leads to a higher disease severity at any time. The negative and

significantly higher than zero value of b_{12} indicates a
negative interaction between the total number of
offspring and the duration of the infectious period
(relative to the latency period). As runs up to T = 50
did not last long enough to exhaust available sites,
the saturation level did not influence this analysis.

Table 5.4 shows that b_1 has the highest absolute
value among the coefficients of function (5.18). Thus,
the outcome $\mathcal{L}$ of the 'diffusion model' is most
sensitive to $\mathcal{R}$. The model's sensitivity to input
parameters decreases in the order $\mathcal{R}$, $\mathcal{S}$ and $\mathcal{U}$. Thus,
special attention must be devoted to experimental
measurements of the number of daughter lesions produced
per mother lesion and to duration of the infectious and
the latency period. The influence of $\mathcal{U}$ was less than
that of $\mathcal{R}$ or $\mathcal{S}$. However, as the runs lasted only 50
simulation days, the position of $\mathcal{U}$ relative to $\mathcal{R}$ and $\mathcal{S}$
may change with higher values of time; as the focus
boundary moves towards the field boundary, more spores
will be lost from the field at higher values of time.

As a second candidate for the 'diffusion model'
response we mentioned the scaled velocity of focus
expansion. Because of the short simulation run time (T
$\leq$ 50), this response could not be determined for runs
with a high value of $\mathcal{U}$; the foci developed only in
close proximity of the point of initial inoculation.

5.6 DISCUSSION

The present method of sensitivity analysis, applied
to computer simulation, allows to evaluate linear,
quadratic and mixed influences of input parameters on
model output. Due to the uniform rotatable central
composite design, the coefficients of the fitted second
order function are determined with equal mean sqare
errors within the desired ranges of the input
parameters. This function gives a good approximation of

the response of the `diffusion model' and thus allows inferences about the influence of the input parameters on the model output.

The results of the sensitivity analysis of the `diffusion theory' can be summarized in a few simple rules:

1. *It is `profitable' for a disease to have a high number of offspring, if the ratio of infectious to latency period is kept constant.*

2. *A low value for the ratio of infectious to latency period leads to a higher lesion number within a field of given size, if the total number of offspring is kept constant.*

3. *The total number of offspring and the ratio of infectious to latency period interact; a high number of offspring together with a short duration of the infectious period lead to higher numbers of lesions than if both factors act separately.*

4. *`Short' dispersal is `profitable' for a disease in the early stages of focus formation, when the effect of exhausting noninfected sites is not yet important.*

These rules are valid for the initial phase of focus formation. They cannot be applied to later stages, when the disease reaches its saturation level in the centre of the focus.

One possible effect of saturation on disease, the optimal partitioning of spores between `short' and `long' dispersal, when two dispersal mechanisms interact, will be discussed in Chapter 9.

6 TELEGRAPHER'S THEORY
OF FOCUS DEVELOPMENT IN PLANT DISEASE

6.1 INTRODUCTION

Chapter 4 has shown that focal epidemics developing within a single field (zero order epidemics sensu Heesterbeek and Zadoks, 1987) can be simulated by numerically solving the equations of the 'diffusion theory'. However, the spatial scale of applicability of this theory is unknown because this theory is based on an idealized picture: During their flight, spores change the direction of their movement an infinite number of times within an arbitrarily short time span (the mean free path for scattering tends to zero and the spore velocity tends to infinity in such a way that the diffusion coefficient is constant). In reality, there is some persistence in the direction of a spore movement. To decide about the spatial scale of applicability of the 'diffusion theory', it should be compared to a theory which assumes some persistence in the direction of a spore's movement. The 'telegrapher's theory', which emerges from replacement of the diffusion equation by the telegrapher's equation, can serve for this purpose, because it assumes that a moving spore has some 'memory' of its direction of flight.

6.2 FOUNDATION OF THE THEORY

6.2.1 Assumptions and definitions

The set of definitions and assumptions underlying the 'telegrapher's theory' is the same as that presented in Section 3.2.1. In Section 3.2.2 the definitions and axioms were put in a phytopathological context, indicating the relation with the phytopathological literature (Plant Pathology Committee, 1950; Anonymous, 1953; Van der Plank, 1963; Vanderplank, 1975).

6.2.2 Terminology

The theory requires its own terminology, which is introduced here, with symbols of variables and parameters (in italics) and their dimensions (in square brackets):

1. x = the first space coordinate [L]
2. y = the second space coordinate [L]
3. z = the third space coordinate [L]
4. t = time [T]
5. $\vec{r}$ = (x,y,z) – a position vector in 3-dimensional space [L]
   Note: The value (length) of the vector $\vec{r}$ will be denoted r; $r = |\vec{r}|$.
6. $\vec{\Omega}$ = a unit vector of a direction in space [1]
7. $\vec{v}$ = velocity of spores $[LT^{-1}]$
   Note: The value (length) of the vector $\vec{v}$ will be denoted v; $v = |\vec{v}|$.
8. s = $s(\vec{r},\vec{\Omega},t)$ – volume density of spores at $\vec{r}$ and t flowing in the $\vec{\Omega}$ direction $[NL^{-3}]$
9. S = $S(\vec{r},t)$ – volume density of spores at $\vec{r}$ and t $[NL^{-3}]$
10. L = $L(\vec{r},t)$ – volume density of lesions at $\vec{r}$ and t $[NL^{-3}]$
11. $\vec{j}$ = $\vec{j}(\vec{r},\vec{\Omega},t)$ – flux at $\vec{r}$ and t of spores flowing in the direction $\vec{\Omega}$; $j = |\vec{j}|$ $[NL^{-2}T^{-1},NL^{-2}T^{-1},NL^{-2}T^{-1}]$
12. $\vec{J}$ = $\vec{J}(\vec{r},t)$ – spore flux at $\vec{r}$ and t; $J = |\vec{J}|$ $[NL^{-2}T^{-1},NL^{-2}T^{-1},NL^{-2}T^{-1}]$

13. C = the macroscopic cross section for a
given process $[L^{-1}]$
Note: The concept of the macroscopic
cross section in 2-dimensional space
was explained in Section 3.3.4. In
3-dimensional space, this
explanation is still valid, after
some reinterpretation due to the
extra spatial dimension.

14. λ_a = the mean free path for absorption;
$\lambda_a = 1/C_a$ where C_a is the
macroscopic cross section for
absorption $[L]$

15. λ_t = the mean free path for
transportation; $\lambda_t = 1/C_t$ where C_t
is the macroscopic cross section for
transportation $[L]$

16. D   = diffusion coefficient $[L^2 T^{-1}]$

6.3 ON THE TELEGRAPHER'S EQUATION

6.3.1 Introduction

The basic equation of the theory to be presented is the telegrapher's equation, which will be derived from the Boltzmann equation. The Boltzmann equation is frequently used in the transport theory of dilute gases and in the theory of the nuclear chain reactor. As indicated in Section 3.3.1, the process of neutron diffusion and multiplication by reaction, and the process of spore diffusion and multiplication by lesions are analogous, but the two processes also differ in many important details. Therefore, the main points of derivation of the 'telegrapher's' equation and its epidemiological interpretation will be discussed here. The derivation follows those by Weinberg and Wigner (1959) and Ash (1979).

6.3.2 Integral formulation – the Boltzmann equation

The continuity equation (3.6) describes the spore flux resulting from a straight-line motion of spores. It can easily be extended to the three-dimensional case. If spores move in the direction $\vec{\Omega}$, equation (3.6) can be rewritten to the direction dependent three-dimensional form:

$$\left[\frac{\partial s(\vec{r},\vec{\Omega},t)}{\partial t}\right]_m = -\nabla \cdot \vec{j}(\vec{r},\vec{\Omega},t) \tag{6.1}$$

where $s(\vec{r},\vec{\Omega},t)$ is the density of spores at $\vec{r}$ and t flowing in the $\vec{\Omega}$ direction, $\vec{j}(\vec{r},\vec{\Omega},t)$ is the flux at $\vec{r}$ and t of spores flowing in the $\vec{\Omega}$ direction, and the subscript m means that equation (6.1) applies to the straight line motion of spores. The following relation links s and $\vec{j}$:

$$\vec{j}(\vec{r},\vec{\Omega},t) = v\,\vec{\Omega}\,s(\vec{r},\vec{\Omega},t) \tag{6.2}$$

where v is the mean velocity of spores.

Apart from a straight-line spore motion, the spore density $s(\vec{r},\vec{\Omega},t)$ changes because of 'scattering' (changes of the direction of spore movement due to air turbulence, collisions with plant surfaces, and so on). The parameter measuring the intensity of this process is called the cross section for scattering (see Section 3.2.4.). The cross section for scattering which changes the direction of spore movement from $\vec{\Omega}'$ to the 'interval' $[\vec{\Omega}, \vec{\Omega}+d\vec{\Omega}]$ (where $d\vec{\Omega}$ is the element of the solid angle $\vec{\Omega}$) will be denoted $C_s(\vec{\Omega},\vec{\Omega}')\,d\vec{\Omega}$. We assume that the cross section is constant all over the field. The rate of change of the density of spores at $\vec{r}$ flowing in the $\vec{\Omega}$ direction due to 'scattering' is expressed by the following equation:

$$\left[\frac{\partial s(\vec{r},\vec{\Omega},t)}{\partial t}\right]_d = \int j(\vec{r},\vec{\Omega}',t)\; C_s(\vec{\Omega},\vec{\Omega}')\; d\vec{\Omega}' -$$

$$\int j(\vec{r},\vec{\Omega},t)\; C_s(\vec{\Omega}',\vec{\Omega})\; d\vec{\Omega}' \qquad (6.3)$$

where the subscript d means that the rate of change of the spore density is due to dispersion. The value $j(\vec{r},\vec{\Omega},t) = |\vec{j}(\vec{r},\vec{\Omega},t)| = v\, s(\vec{r},\vec{\Omega},t)$. The integrals are calculated over all directions (full range of $\vec{\Omega}'$). The first term at the right hand side of (6.3) describes the 'scattering' events, which change the direction to $\vec{\Omega}$ from any arbitrary direction $\vec{\Omega}'$. The second term describes the 'scattering' events, which change the direction from $\vec{\Omega}$ to any arbitrary direction $\vec{\Omega}'$. In common words, equation (6.3) means:

The rate of change of the density of spores moving in the $\vec{\Omega}$ direction equals the sum of all spore density currents having moved previously in the $\vec{\Omega}'$ direction (where $\vec{\Omega}'$ is an arbitrary direction different from $\vec{\Omega}$) which changed their direction to $\vec{\Omega}$, minus the sum of the spore density currents traveling previously in $\vec{\Omega}$ direction which changed their direction to an arbitrary $\vec{\Omega}'$ direction.

Any 'scattering' event from $\vec{\Omega}$ changes the direction of a spore and removes it from $j(\vec{r},\vec{\Omega},t)$, so that the second term of the right-hand side of (6.3) can be written as $j(\vec{r},\vec{\Omega},t)\, C(\vec{\Omega})$, where

$$C_s(\vec{\Omega}) = \int C_s(\vec{\Omega}',\vec{\Omega})\; d\vec{\Omega}' \qquad (6.4)$$

is the cross section for arbitrary 'scattering' from the $\vec{\Omega}$ direction.

Another cause of changes in the spore density is spore deposition (absorption). The rate of change of the spore density due to deposition is:

$$\left[\frac{\partial s(\vec{r},\vec{\Omega},t)}{\partial t}\right]_a = -\,\delta\, s(\vec{r},\vec{\Omega},t) \qquad\qquad (6.5)$$

where δ is the rate of spore deposition and the subscript a means that the rate of change of the spore density is due to absorption. The minus sign means that the deposition decreases the spore density.

In addition to straight-line motion, `scattering' and deposition, a change in the density of spores in $\vec{r}$ and t moving in the $\vec{\Omega}$ direction results from the production of spores by sporulating lesions. The production term, giving input of spores in $\vec{r}$ and $\vec{\Omega}$, will be denoted as $P_\Omega = P_\Omega(\vec{r},\vec{\Omega},t)$.

Finally, inserting (6.1), (6.3), (6.5) and the production term into the spore balance equation (3.1), the total rate of change of the spore density is:

$$\frac{\partial s(\vec{r},\vec{\Omega},t)}{\partial t} = -\,\nabla\cdot\vec{j}(\vec{r},\vec{\Omega},t) - \delta\, s(\vec{r},\vec{\Omega},t) -$$

$$C_s(\vec{\Omega})\, j(\vec{r},\vec{\Omega},t) + \int j(\vec{r},\vec{\Omega}',t)\, C_s(\vec{\Omega},\vec{\Omega}')\, d\vec{\Omega}' +$$

$$P_\Omega(\vec{r},\vec{\Omega},t) \qquad\qquad (6.6)$$

where relation (6.4) was applied to the second integral from (6.3). Equation (6.6) is known in physics as the Boltzmann equation. It is the mathematical formulation of the following sentence:

The rate of change of the density in $\vec{r}$ at t of spores, which flow in the $\vec{\Omega}$ direction, is the sum of the contributions of five processes:

1. *straight-line motion in the $\vec{\Omega}$ direction (the first term),*
2. *absorption (the second term),*
3. *scattering from the $\vec{\Omega}$ direction to every other $\vec{\Omega}'$ direction (the third term),*
4. *scattering to the $\vec{\Omega}$ direction from every other direction $\vec{\Omega}'$ (the fourth term), and*
5. *production of spores (the fifth term).*

6.3.3 Differential formulation
– the telegrapher's equation

The Boltzmann equation is the general equation
describing the `behaviour' of spores in a `scattering'
(turbulent), `absorbing' (deposition), and multiplying
(a single spore infects a site, which then produces
many spores) medium such as a crop canopy. The equation
is unwieldy. For numerical analysis, its approximate,
differential form - the `telegrapher's' equation - is
more convenient. This equation will be derived below,
but only the main points of the derivation will be
presented. A more detailed derivation can be found in
Ash (1979) and Weinberg and Wigner (1959).

Multiplying equation (6.6) by $d\vec{\Omega}$, integrating over
all solid angles, and noting that:

$$\int C_s(\vec{\Omega})\ j(\vec{r},\vec{\Omega},t)\ d\vec{\Omega} = \int \left[\int C_s(\vec{\Omega},\vec{\Omega}')\ d\vec{\Omega} \right] j(\vec{r},\vec{\Omega}',t)\ d\vec{\Omega}'$$

$$= \int \left[\int C_s(\vec{\Omega}',\vec{\Omega})\ d\vec{\Omega}' \right] j(\vec{r},\vec{\Omega},t)\ d\vec{\Omega}$$

leads to:

$$\frac{\partial S(\vec{r},t)}{\partial t} + \nabla\cdot\vec{J}(\vec{r},t) + \delta\ S(\vec{r},t) = P \qquad (6.7)$$

where:

$$S(\vec{r},t) = \int s(\vec{r},\vec{\Omega},t)\ d\vec{\Omega}$$

$$\vec{J}(\vec{r},t) = \int j(\vec{r},\vec{\Omega},t)\ \vec{\Omega}\cdot d\vec{\Omega} = \int \vec{j}(\vec{r},\vec{\Omega},t)\ d\vec{\Omega}$$

$$P(\vec{r},t) = \int P_\Omega(\vec{r},\vec{\Omega},t)\ d\vec{\Omega}$$

and $P_\Omega(\vec{r},\vec{\Omega},t)$ is assumed to be isotropic. Equation

(6.7) is the continuity equation formulated in the presence of absorption and production. It should be read:

The rate of change of the spore density at $\vec{r}$ and t (dS/dt) is equal to the rate of spore production (P) minus the net rate of spore outflow ($\nabla\ \vec{J}(\vec{r},t)$) minus the rate of spore deposition (absorption) ($\delta\ S(\vec{r},t)$).

Multiplying equation (6.6) by $\vec{\Omega}\ d\vec{\Omega}$ and integrating over all solid angles and assuming that $C_s(\vec{\Omega}) = C_s$ (C_s is independant of $\vec{\Omega}$; scattering is the same from every direction) leads to:

$$\frac{\partial S(\vec{r},t)}{\partial t}\ \bar{\vec{\Omega}} + \int \nabla\cdot\vec{j}(\vec{r},\vec{\Omega},t)\ \vec{\Omega}\ d\vec{\Omega} + \delta\ S(\vec{r},t)\ \bar{\vec{\Omega}} +$$

$$C_s\ \vec{J}(\vec{r},t) = \iint j(\vec{r},\vec{\Omega}',t)\ C_s(\vec{\Omega},\vec{\Omega}')\ d\vec{\Omega}'\ \vec{\Omega}\ d\vec{\Omega} \qquad (6.8)$$

where

$$\bar{\vec{\Omega}}(\vec{r},t) = \left[\int s(\vec{r},\vec{\Omega},t)\ \vec{\Omega}\ d\vec{\Omega}\right] / S(\vec{r},t)$$

Further analysis needs the assumption that $j(\vec{r},\vec{\Omega},t)$ and $C_s(\vec{\Omega},\vec{\Omega}')$ are almost isotropic, so that they can be expanded in Legendre polynomials. This assumption implies that:

1. the volume of interest is many mean free paths away from anisotropic sources or sinks (e.g. the boundary of the field from which spores cannot be scattered back into the field),

2. almost isotropic diffusion is the main phenomenon occurring. In other words, the spore density current is almost isotropic and its magnitude changes only slightly within one mean free path for scattering.

Under the assumptions stated above, the spore density current and the cross section for scattering can be expanded in Legendre polynomials. Retaining only the first two terms of these expansions, and performing some mathematical operations (see Ash, 1979, pages 13-15), leads to:

$$\frac{\partial S(\vec{r},t)}{\partial t}\,\hat{\Omega} + C_t\,\vec{J}(\vec{r},t) + \frac{1}{3}\,\nabla\,J(\vec{r},t) = 0 \qquad (6.9)$$

where $C_t = C_s\,(1 - \mu)$ (μ is the average cosine of the scattering angle) is the transportation cross section (Weinberg and Wigner, 1959; Ash, 1979). In the case of spore dispersal C_t is an empirical parameter. The inverse of C_t, the mean free path for transportation, λ_t, can be measured in the same way as its 'diffusion' analog, the mean free path for scattering or the 'mixing length' (Goudriaan, 1977). Equation (6.9) was derived from (6.8).

Using (6.9), the term $\nabla \cdot \vec{J}(\vec{r},t)$ can be eliminated from (6.7) (see Weinberg and Wigner, 1959, page 235). Finally applying the definition of $J(\vec{r},t)$, this leads to:

$$\frac{3 \cdot D}{v^2}\,\frac{\partial^2 S(\vec{r},t)}{\partial t^2} + \left[1 + \frac{3 \cdot D \cdot \delta}{v^2}\right]\,\frac{\partial S(\vec{r},t)}{\partial t} =$$

$$D\,\nabla^2\,S(\vec{r},t) - \delta\,S(\vec{r},t) + P \qquad (6.10)$$

where the diffusion coefficient $D = v\,\lambda_t/3$, and the mean free path for transportation $\lambda_t = 1/C_t$, and ∇^2 is the differential operator

$$\nabla^2 = \frac{\partial^2}{\partial x} + \frac{\partial^2}{\partial y} + \frac{\partial^2}{\partial z} \qquad (6.11)$$

Equation (6.10) is called the 'telegrapher's' equation. It is a combination of the wave and the diffusion equation. This equation can be explained as follows:

Spores are dispersed like a dissipating wave; the first term of the left-hand side and the first term of the right-hand side are the wave equation. However after passing of the wave front, a 'residual disturbance' (spores still air-borne) due to the diffusion equation remains; the second term of the left-hand side and the first term of the right-hand side are the diffusion equation. In addition, spores are deposited (the second term of the right-hand side) and produced (the third term of the right-hand side).

The rate of spore deposition can be interpreted in the scattering model (see Chapter 3) as $\delta = v \lambda_a$, where λ_a is the mean free path for absorption (the distance of spore displacement at which $1/e$ spores is not yet absorbed).

Translation to the two-dimensional space can be done by reinterpretation of S, replacing the factor 3 by 2 in equation (6.10) and in the definition of the diffusion coefficient, and by replacing (6.11) by its two-dimensional counterpart:

$$\nabla^2 = \frac{\partial^2}{\partial x} + \frac{\partial^2}{\partial y} \tag{6.12}$$

The solution of the 'telegrapher's' equation shows the phenomenon of 'retardation', i.e. the solution at time t takes non-zero values only for those points of space which are less distant from the origin of the focus than the distance travelled by a spore with velocity v during t. In addition to 'residual disturbance' (non-zero value of the density of spores S), which persists at all points passed by the wave front, the 'telegrapher's' equation has a well-defined front. In the region beyond the distance travelled by a spore during the time span considered, the solution of equation (6.10) is zero.

Letting the spore velocity v grow without limit, and

the mean free path for transportation tend to zero in
such a way that D remains finite, and the mean free
path for absorption tends to infinity in such a way
that $\delta = v/\Lambda_a$ remains finite, the asymptote to equation
(6.10) is obtained:

$$\frac{\partial S(\vec{r},t)}{\partial t} = D \, \nabla^2 \, S(\vec{r},t) - \frac{v}{\Lambda_a} \, S(\vec{r},t) + P \qquad (6.13)$$

Equation (6.13) is the diffusion equation (3.28)
introduced in Chapter 3.

6.4 EQUATIONS FOR FOCUS DEVELOPMENT

Solutions of equation (6.10) depend on the
production term P. In Chapter 3, the general form of
this term, equation (3.39), was derived. The equation
describing the time dependency of production of new
lesions by spores was presented above as equation
(3.45).

Substitution of (3.39) into equation (6.10) leads to
the following telegrapher's equation:

$$\frac{A \cdot D}{v^2} \frac{\partial^2 S(\vec{r},t)}{\partial t^2} + \left[1 + \frac{A \cdot D \cdot \delta}{v^2} \right] \frac{\partial S(\vec{r},t)}{\partial t} =$$

$$D \, \nabla^2 \, S(\vec{r},t) - \delta \, S(\vec{r},t) + \int_0^\infty \frac{\partial L(\vec{r},t-\tau)}{\partial t} \, K(\tau) \, d\tau \qquad (6.14)$$

In two-dimensional space $A = 2$, the diffusion
coefficient $D = v \, \lambda_t/2$ and ∇^2 is the differential
operator (6.12). In three-dimensional space $A = 3$, $D =
v \, \lambda_t/3$ and ∇^2 is the differential operator (6.11).

Equations (6.14) and (3.46) constitute the system of
partial differential equations, for spores and lesions

respectively, which are the mathematical formulation of the `telegrapher's theory' of focus development in time and space.

The system described above is too complex for an analytical solution to be used in practical applications. A numerical solution is needed, for which the computer package PODESS was written (Appendix A).

6.5 A NUMERICAL COMPARISON TO THE `DIFFUSION' THEORY

The `diffusion' (Chapter 3) and the `telegrapher's' theories attempt to describe focus development. The `diffusion theory', based on the diffusion equation, is the asymptote to the `telegrapher's theory', based on the telegrapher's equation. The equations of the `telegrapher's theory' are more complex (the telegrapher's equation is second order in time) than the equations of the `diffusion theory' (the diffusion equation is first order in time), so that they need more computer time and more computer memory for their solution. Therefore, it is preferable to use the `diffusion theory' whenever its results are consistent with the results of the `telegrapher's theory'.

The parameter playing a crucial role in the decision which theory should be used is the spore velocity. The following numerical example compares the results of the two theories in dependence on the spore velocity.

Results were obtained by six runs of the computer program PODESS (Appendix A), three runs for the `diffusion model' and three for the `telegrapher's model'. The independent variable, spore velocity, was given the values 0.02 m/day, 0.2 m/day, and 2 m/day. Theoretically, the ratio of this velocity to the diameter of the `solution' region multiplied by the time-period of interest decides which theory gives the best approximation of reality. The period of simulation between two succesive outputs (so called `communication interval') was one 'simulated' day, so that this period was used as the time-period of interest. The diameter of the 'simulated' field was 10 m x 10 m. As the initial conditions, (1) one lesion at the centre of the field, (2) no spores within the field, and (3) zero value of the first time derivative of the spore density, were chosen.

Table 6.1. A comparison between the 'diffusion theory' and the 'telegrapher's theory'. Results of the two theories in dependence of the spore velocity.
† Distance from the point of initial inoculation.
‡ The lesion density (number of lesions per grid point; one grid point represents an area of 0.5 m x 0.5 m).

Spore veloc. in m/day	Dist. in m†	'Diffusion' theory result‡	'Telegrapher's' theory result‡	Difference in %
0.02	0.	1852.	1770.	4.6
	0.5	33.	30.	9.2
	1.	0.36	0.32	12.5
	1.5	0.0029	0.0024	20.8
	2.	0.000019	0.000014	35.7
0.2	0.	271640.	272650.	-0.4
	0.5	30778.	30790.	-0.04
	1.	1673.	1664.	0.5
	1.5	64.3	63.6	1.1
	2.	1.98	1.95	1.5
2.	0.	1086800.	1089500.	-0.25
	0.5	606150.	614180.	-1.3
	1.	95202.	97405.	-2.3
	1.5	7474.	7669.	-2.6
	2.	445.	458.	-2.7

The results (Table 6.1) show no great differences between the two theories. The difference tends to decrease with increasing value of the spore velocity.

The increase in the differences for the spore velocity of 2 m/day relative to those for the spore velocity of 0.2 m/day seems to be an effect of the single precision calculations, which are not always sufficiently accurate. For high values of the spore velocity the equations become `stiff' (Gear, 1971); they require shorter time steps of integration, and thus lead to an accumulation of so called `rounding-off errors'. If a higher accuracy is required, double precision calculations should be used.

6.6 DISCUSSION

The `telegrapher's theory' of focal disease development results from a picture of the reality of spore dispersal different from that of the `diffusion theory'. The `telegrapher's theory' resulting from the underlying simplified picture of spore dispersal (spores move along straight lines and suddenly change the direction of their movement by scattering), also has its limitations. They result from the assumptions made in the derivation of the telegrapher's equation. The first assumption (the volume of interest is many mean free paths away from anisotropic sources or sinks; Section 6.3.3), may lead to some trouble near the boundary of the `solution' region and near the field boundaries (if the `solution' region contains separate fields). The second assumption (almost isotropic diffusion is the main phenomenon occurring) is not a very rigid restriction in the case of air-borne fungal spores, because their mean free path for transportation is a few centimetres to a few decimetres long (Goudriaan, 1977), whereas the mean free path for absorption, derived from the variance of the Bessel distribution of lesions (Williams, 1961; Van den Bosch et al., 1988 b) is usually many times longer.

The numerical analysis of the system of equations (6.14), (3.55) is more difficult than the analysis of the system of equations constituting the mathematical formulation of the 'diffusion theory'. Equation (6.14) is a second order, in the time derivative, partial differential equation. To treat this equation by PODESS, it must be decomposed into a system of two first order, in the time derivative, partial differential equations. Together with equation (3.46) these equations constitute the system of three partial differential equations, which consume more computer time and memory, during a solution run, than the system of 'diffusion theory' equations. The comparison of the numerical results of the 'diffusion' and the 'telegrapher's' theories helps to answer, if and when the 'diffusion theory', is adequate to describe focal disease development.

The 'telegrapher's theory' exhibits an important conceptual consequence, a well-defined front of the disease (the border-line between the regions where $S(\vec{r},t) > 0$ and $S(\vec{r},t) = 0$). However, this front of the disease moves with a velocity which is much larger than the velocity of movement of the contours of constant non-zero severity movement. So it is not of much use in practice.

From a theoretical point of view, the 'diffusion theory' is the limit to the 'telegrapher's theory', based on the following assumptions: (1) the spore velocity tends to infinity and the mean free path for transportation tends to zero in such a way, that their product is constant, (2) the mean free path for absorption tends to infinity in such a way that its ratio to the spore velocity is constant. These assumptions lead to the hypothesis, that the difference between the results of the two theories should decrease with growing value of the spore velocity. This conclusion was confirmed by calculations, but the differences were never large, even far from the limiting situation.

It may be concluded that the `diffusion theory' will provide adequate results, at least for qualitative analysis, even when the spore velocity is relatively low. Thus, computer time can be saved by using the `diffusion theory'.

7 SIMULATION OF WIND EFFECTS ON FOCUS DEVELOPMENT

7.1 INTRODUCTION

Many authors tried to explain the effect of wind on the development of epidemics. Fundamental work was done by Schroedter (1960). Okubo (1980) presented, as an example, an elliptical shape of the region with disease, if there was a prevailing wind. Zadoks et al. (1969) described an experiment with smoke pufs, seen as models of spore clouds, and discussed their behaviour. Gregory (1968) determined gradients downwind and upwind of a source of infection. In his 1973 book, Gregory also discussed dispersal under the influence of wind. The effect of wind on deposition of *Lycopodium* spores in wind tunnel experiments was shown by Chamberlain and Chadwick (1972). Studies on the influence of wind gusts on particle liberation and deposition were made by Aylor (Aylor, 1987; Aylor et al., 1981). Pedgley (1982) and Jeger (1985a, b) gave many examples of the long-range spore transport by wind. Though the literature on wind and plant disease is vast, few papers refer specifically to the effect of wind on focus development.

The objective of this chapter is to generalize the diffusion equation to situations of focus development under the influence of wind.

7.2 THE STRANGE WORLD OF MOVING SYSTEMS

For a few hundred years, systems with moving coordinates have been used in physics. They help to solve complicated problems in a comparatively simple way, if we only know the rules of behaviour of a group of objects in one coordinate system (often the best system is the `resting system' in which a group of objects as a whole is at rest). Only a short description of moving coordinate systems will be given here. More information can be found in handbooks of physics (see Kittel et al., 1965).

If one coordinate system (x,y,z,t), where x,y,z describes a space and t measures a time, is at rest for the observer and another system (x',y',z',t'), where x',y',z' describes a space and t' measures a time, moves with velocity $\vec{w}$ which is constant in value and direction, then such systems are called `inertial'. According to the Galilei principle, all physical laws are equal in inertial systems, after `Galilean transformation' of the coordinates. The mathematical form of this transformation is:

$$t' = t$$
$$x' = x - w_x t$$
$$y' = y - w_y t$$
$$z' = z - w_z t \qquad\qquad (7.1)$$

where x, y, z, and t are the space-time coordinates of the resting system, x', y', z', and t' are the space-time coordinates of the system which moves with constant velocity $\vec{w}$ and w_x, w_y, w_z are the x-, y-, and z-components of $\vec{w}$, respectively. The last three lines can be expressed together by

$$\vec{r}' = \vec{r} - \vec{w} t$$

where $\vec{r}' = (x',y',z')$ and $\vec{r} = (x,y,z)$. The transformation is visualized in Fig 7.1, where the only non-zero component of $\vec{w}$ is w_x.

If we know the law governing a process in the resting system, we can describe the same process in a moving system by changing the coordinates according to (7.1).

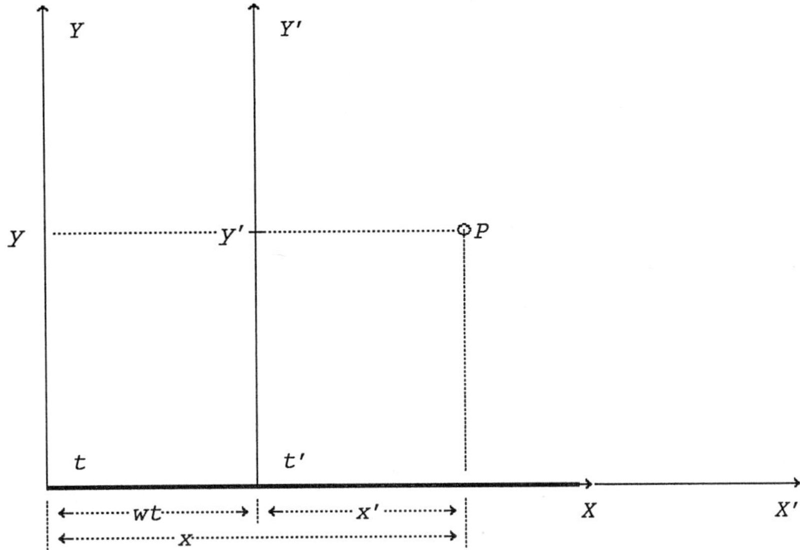

Fig. 7.1. Two-dimensional picture to demonstrate the Galilean transformation in the x-direction. Directions z and z' are perpendicular to the surface of the paper. They can be treated like y and y', because the direction of movement is perpendicular to y, y' and z, z'. In the resting system, the coordinates of a point P are x, y, t (t is the current time in the resting system) and in the moving system they are $x' = x - w\,t$, y', t' (t' is the current time in the moving system). Because P is arbitrary, the transformation is valid for every point.

Conversely, if the law is given for a moving system and we want to know the law for a resting system, we use the inverse transformation:

$$t = t'$$
$$x = x' + w_x\, t'$$
$$y = y' + w_y\, t'$$
$$z = z' + w_z\, t' \tag{7.2}$$

7.3 WIND AND THE DIFFUSION EQUATION

If the wind blows with velocity $\vec{w}$, the whole air mass and everything in it moves with the velocity $\vec{w}$ in the direction of the wind, but the spore sources stay put. The air eddies (air turbulences) move with velocity $\vec{w}$. In the resting system of the wind (the system in which the air mass as a whole does not move), the system of equations (3.45) and (3.46) describes disease development.

The foregoing can be visualized by means of the following analogy. In a train, moving at a velocity of 100 km/h, the passenger can describe the route followed by a fly in his compartment taking into consideration only the velocity of the fly relative to the walls of his carriage. However, for an observer standing on a platform of a railway station, the velocity of the fly is one hundred kilometers per hour higher.

Standing in a wheat field, we are in the situation of the observer on the platform. If we want to describe disease development in a three-dimensional space, including wind effects, we should use the three-dimensional analog of equation (3.46), replacing x, y, z and t by x', y', z' and t', respectively (because the system, in which the equation is valid, moves relative to the field) and apply transformation (7.1). Equation (3.45) should be left unchanged because the wind does not influence lesion formation.

As densities do not depend on a coordinate system in which they are measured

$$S'(\vec{r}',t') = S(\vec{r},t)$$
$$L'(\vec{r}',t') = L(\vec{r},t) \tag{7.3}$$

where $S'(\vec{r}',t')$ and $S(\vec{r},t)$ are the spore densities expressed as functions of the moving and the resting coordinates, and $L'(\vec{r}',t')$ and $L(\vec{r},t)$ are the lesion densities expressed as functions of the moving and the resting coordinates, respectively. As the production term is expressed in numbers per day, the same relation applies to it:

$$P'(\vec{r}',t') = P(\vec{r},t) \tag{7.4}$$

where $P'(\vec{r}',t')$ is the production term expressed as function of the moving coordinates and $P(\vec{r},t)$ is the production term expressed as function of the resting coordinates.

From the resting system point of view, the values of space coordinates in the moving system are t dependent (see (7.1)). Therefore, the partial derivative of the spore density, with respect to time, must be replaced by the so called material derivative, which can be expressed in the resting system by:

$$\frac{d}{dt'} = \frac{\partial}{\partial t}\frac{\partial t}{\partial t'} + \frac{\partial}{\partial x}\frac{\partial x}{\partial t'} + \frac{\partial}{\partial y}\frac{\partial y}{\partial t'} + \frac{\partial}{\partial z}\frac{\partial z}{\partial t'} =$$

$$\frac{\partial}{\partial t} + w_x\frac{\partial}{\partial x} + w_y\frac{\partial}{\partial y} + w_z\frac{\partial}{\partial z} \tag{7.5}$$

where transformation (7.2) was applied. Therefore,

$$\frac{dS'(\vec{r}',t')}{dt'} = \frac{\partial S(\vec{r},t)}{\partial t} + w_x\frac{\partial S(\vec{r},t)}{\partial x} + w_y\frac{\partial S(\vec{r},t)}{\partial y} +$$

$$w_z\frac{\partial S(\vec{r},t)}{\partial z} \tag{7.6}$$

where relation (7.3) and (7.5) were used.

For space derivatives:

$$\frac{\partial^2}{\partial x'^2} = \frac{\partial^2}{\partial x^2}$$

$$\frac{\partial^2}{\partial y'^2} = \frac{\partial^2}{\partial y^2}$$

$$\frac{\partial^2}{\partial z'^2} = \frac{\partial^2}{\partial z^2} \qquad (7.7)$$

Finally, we apply (7.3), (7.4), (7.6) and (7.7) to the diffusion equation, formulated in the moving coordinate system, in order to translate it to the resting system. The new equation including the effect of the wind is:

$$\frac{\partial S(\vec{r},t)}{\partial t} = D \left[\frac{\partial^2}{\partial x^2} + \frac{\partial^2}{\partial y^2} + \frac{\partial^2}{\partial z^2} \right] S(\vec{r},t) -$$

$$w_x \frac{\partial S(\vec{r},t)}{\partial x} - w_y \frac{\partial S(\vec{r},t)}{\partial y} - w_z \frac{\partial S(\vec{r},t)}{\partial z} -$$

$$\delta\, S(\vec{r},t) + \int_0^\infty \frac{\partial L(\vec{r},t-\tau)}{\partial t} K(\tau)\, d\tau \qquad (7.8)$$

$$\frac{\partial L(\vec{r},t)}{\partial t} = E \left[1 - \frac{L(\vec{r},t)}{L_{max}} \right] \delta\, S(\vec{r},t) \qquad (7.9)$$

The system of equations (7.8), (7.9) generalizes the system (3.46), (3.45), including the wind effect on disease development.

By varying w_x, w_y and w_z we can use the system of equations (7.8) and (7.9) to describe focus development as affected by a wind variable in time. In the field situation, when the wind changes all the time, this is the most interesting case.

In the case of a constant wind, asymptotic focus

shape and asymptotic velocity of focus extension can be
obtained analytically in the framework of an extension
of the Diekmann-Thieme theory (Van den Bosch et al., in
prep.).

As in previous cases, the system of equations (7.8),
(7.9) can be treated numerically by PODESS (see
Appendix A).

7.4 LEAF RUST ON WHEAT - THE TWO-DIMENSIONAL CASE

The results obtained by the numerical solution of
the `diffusion model', based on the `diffusion theory',
were extended by adding a wind effect. The results were
compared to experimental data. The experiment, designed
according to the microfield technique (Zadoks and
Schein, 1979), was performed in 1986, with 10 identical
plots 1.8 x 1.8 m where 7 x 7 hassocks of a susceptible
wheat (cv. Marksman) were planted. The centre hassocks
were inoculated in a greenhouse with leaf rust
(*Puccinia recondita* f.sp. *tritici*) and then transfered
to the field at 21 April. Measurements were made four
times in July: 1st, 9th, 17th, and 23rd. Severity per
hassock for each of three leaf layers was assessed
according to the Peterson B-scale (Zadoks and Schein,
1979). The means of these measurements (Fig. 7.2C) were
used for comparison with the numerical results of the
`diffusion model'.

7.4.1 Parameters

Unfortunately, no parameters required by the
`diffusion theory' were measured. Therefore some
parameters values were estimated from data available in
the literature, the others were guesstimated. Focus
development was simulated with the following parameter
values:

1. D - diffusion coefficient = 0.03 [m^2/day],

2. δ - rate of spore deposition = 0.7 [1/day],
3. E - effectiveness = 1,
4. R - productivity = 3.5 [number of daughter lesions
    per mother lesion per day],
5. p - latency period = 22 [days],
6. i - infectious period = 15 [days],
7. A - area of a single lesion = 10 [mm^2],
8. LAI - leaf area index = 1.5,
9. w - the velocity of wind is given in Table 7.1; for
   the intermediate days, the wind velocity was
   calculated by linear interpolation from the two
   nearest dates.

Ad 8. The value of LAI is the effective leaf area index
   available for infection (the experimental data
   were measured in Peterson B-scale for which 100%
   severity is given to 37% of infected leaf area;
   see Zadoks and Schein, 1979)

The crop was inoculated at the centre of the field at
time T = 0 by 1 successful spore.

Table 7.1 The velocity of wind (in m/day) used for the
simulation of the leaf rust focus development on wheat.
The values are very low, but they represent the mean
wind speed per day.

Day number	Wind in the x-direction	Wind in the y-direction
0	0.	0.06
70	0.	0.06
71	-0.1	0.06
75	0.1	0.06
79	0.2	0.06
100	0.2	0.06

7.4.2 Results

Results were obtained by two runs of the `diffusion model' on a VAX 785 computer, using PODESS (Appendix A), simulating focus development in two-dimensional space (the two-dimensional analog of the system (7.8), (7.9) was used) during 100 days. The field was represented by a grid of 19 x 19 points. The first run was a `control' and it was performed by the `diffusion model' based on the `diffusion theory' presented in Chapter 3 (without wind). In the second run, the `diffusion model' used the extended version of the `diffusion theory' with the wind effect included.

Results of the first run (Fig. 7.2A) show a circular focus developing around the centre of the field. Results of the second run (Fig. 7.2B) are qualitatively different. The focus is an ellipse rather than a circle and its centre moved away from the field centre.

7.4.3 Discussion

Comparison of the numerical results to those obtained experimentally shows that the `diffusion model' without simulation of the wind effect is inadequate in the present case. Results of the `diffusion model' based on the extended version of the `diffusion theory' allow to simulate the directional effect of wind on focus development. However, quantitative differences were observed between the experimental and the numerical results obtained by the second run. They are probably due to wrong parameter guestimates. Application of more detailed models of computer simulation may improve these results. Chapter 8 discusses some of the possible methods and Chapter 9 presents some models. One of them (Section 9.4) shows how more realistic results can be obtained for the case discussed above.

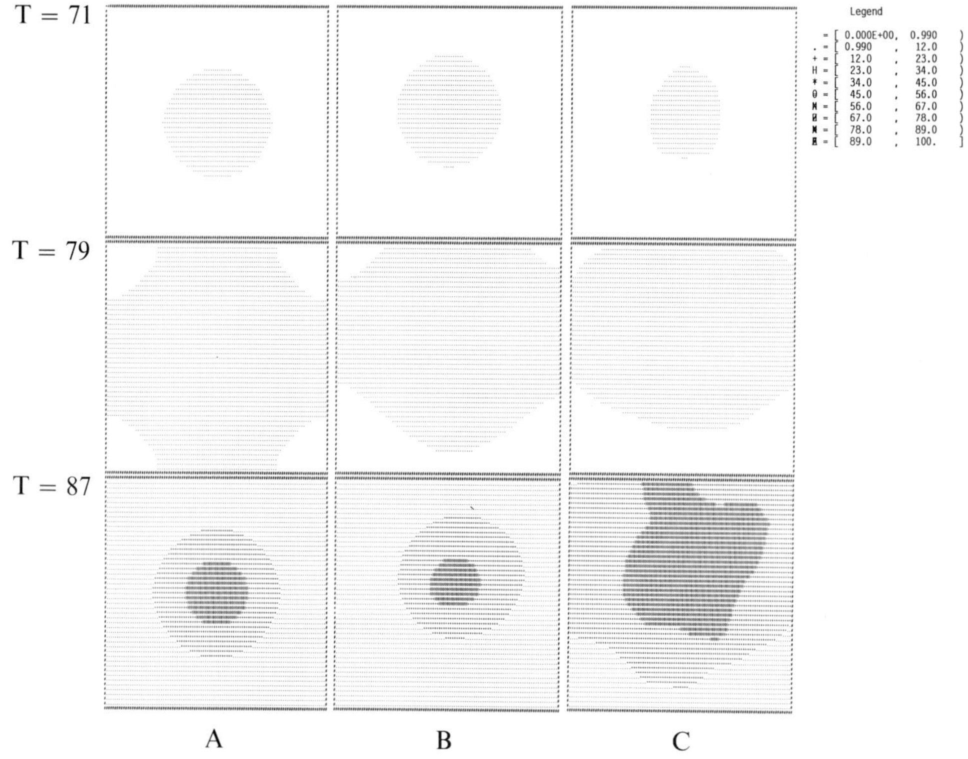

Fig. 7.2. Focus development under the wind influence for time T = 71, 79 and 87. X- and Y-axes are distances from 0 to 1.8 m. **A,** simulation of focus development without wind. **B,** simulation of focus development with wind which velocity is given in Table 7.1. **C,** experimental results of the leaf rust focus development on wheat.

8 A SIMULATION APPROACH

8.1 INTRODUCTION

The `diffusion theory' combines elementary physical and phytopathological concepts into a set of equations, which can be solved numerically using a computer. The structured set of equations, ready for computer use, is called the `diffusion model'. The results of this model can be compared with results of other models and - more important - results of experiments. By comparison, the `diffusion model' can be validated (Chapter 4). The `diffusion model' can be solved numerically by application of appropriate software, in the present case PODESS (Appendix A). The flexibility of PODESS allows not only a numerical solution in a simple case (uniform crop, constant parameter values), but also simulation of more complex situations. Complicated cropping patterns and variable environmental conditions can be handled. Thus, the `diffusion model' becomes a sophisticated tool in the hands of phytopathologists for every-day work rather than just one more model, which may be theoretically justified but is of limited practical value. In this perspective the `diffusion theory' becomes a framework within which a variety of specialized applications can be developed. A few examples will be given in this chapter.

8.2 PARAMETERS AS EMPIRICAL FUNCTIONS

Only a theoretician, in his simplified view of reality, can assume that the model parameters are constant. In practice, they always vary, both deterministically in dependence on known external conditions and stochastically. Any model attempting to

reflect nature, must take this parameter variability
into consideration. Parameter stochasticity often can
be disregarded when looking for qualitative consistency
of the 'diffusion model' with the experimental data.
Then the mean value of a parameter can be used. But
functional dependence of parameter values on external
conditions must be simulated. As we cannot describe the
complete environment, a system containing only the most
relevant elements must be chosen and modelled (de Wit
and Goudriaan, 1978; de Wit, 1982). Influence on the
system from outside can be incorporated into the model
by treating parameters as empirical functions of
external conditions. Thus, the 'diffusion model' can
simulate the effects of changes in weather (by
functional dependence of the diffusion coefficient or
the spore production rate on temperature, humidity,
wind, etc.), in crop growth stages (changes in the
probability of infection with time, etc.), or in
spatial distribution of crop elements (changes of
parameter values with position). Spatial variation of D
should be treated with special care as then D cannot be
put outside the space derivatives in the diffusion
equation. The way to overcome this problem is
analogical to the one proposed for the
three-dimensional case in Section 8.3.2.

8.3 VERTICAL SPORE DISTRIBUTION –
THE THIRD SPATIAL DIMENSION

8.3.1 Introduction

The 'diffusion model' for time and two-dimensional
space was derived in Chapter 3 and validated in Chapter
4. Good qualitative and in most cases good quantitative
consistency with other models and with experimental
data was achieved. But a real crop is three-dimensional
in space. There exist: (1) a soil, on which spores are

deposited, (2) a host canopy, which can be infected by deposited spores and then carry sporulating lesions, and (3) an air layer above a crop, within which spores are transported. Under gravitation, spores fall down continuously. `Stokes' law describes this fall as a downward movement with constant velocity, usually called the `settling velocity' (Okubo, 1980).

8.3.2 The `diffusion theory'
in three-dimensional space

The `diffusion theory' formulated for time and two-dimensional space (Chapters 3 and 7) can be extended to the three-dimensional case, with a uniform vertical crop distribution. To upgrade the theory, a three-dimensional interpretation must be given to the density parameters. The system of equations (7.13), (7.14) must be replaced by its three-dimensional version. Equation (7.13) was derived to describe a wind effect, but in fact it applies to any situation with a prevailing direction of spore movement; here the downward direction, simulating gravitation. The new system formulates the basis of the three-dimensional `diffusion theory' for a uniform crop:

$$\frac{\partial S(\vec{r},t)}{\partial t} = D_C \left[\frac{\partial^2}{\partial x^2} + \frac{\partial^2}{\partial y^2} + \frac{\partial^2}{\partial z^2} \right] S(\vec{r},t) +$$

$$v_g \frac{\partial S(\vec{r},t)}{\partial z} - \delta_C S(\vec{r},t) +$$

$$\int_0^{\infty} \frac{\partial L(\vec{r},t-\tau)}{\partial \tau} K(\tau) \, d\tau \tag{8.1}$$

$$\frac{\partial L(\vec{r},t)}{\partial t} = E \left[1 - \frac{L(\vec{r},t)}{L_{max}} \right] \delta_C S(\vec{r},t) \tag{8.2}$$

where $\vec{r} = (x,y,z)$ represents a position in three-dimensional space, t stands for time, $S(\vec{r},t)$ is the spore density, $L(\vec{r},t)$ is the lesion density, D_c is the diffusion coefficient within the crop, δ_c is the rate of spore deposition within the crop, and v_g is the 'settling velocity' (the downward spore velocity resulting from gravitation and the resistive force of the air). D_c and δ_c are assumed to be constant.

In reality, the vertical crop distribution is not uniform. Therefore, the system of equations (8.1), (8.2) applies only to appropriate parts of the three-dimensional space which contains the crop. Spores are deposited there, and a fraction of them initializes lesions, which produce new spores. Because of slicing this part of three-dimensional space to represent leaf layers (Goudriaan, 1977; Zadoks and Schein, 1979), equation (8.2) must be discretized by establishing a system of as many two-dimensional equations, analogous to (8.2), as there are leaf layers.

For the soil layer, where the only process is spore deposition, the system (8.1), (8.2) is replaced by the absorbing boundary condition (the spore density equals zero); see Chamberlain (1967), Chamberlain and Chadwick (1972) and Aylor et al. (1981) for more information about spore deposition on the soil.

Apart from a crop and a soil, the third element of three-dimensional reality is the air layer above a crop. Spores can be transported there by diffusion and gravitation, but they are not deposited. There are neither lesions nor spore production. The system of equations (8.1), (8.2) is replaced by the following equation:

$$\frac{\partial S(\vec{r},t)}{\partial t} = D_a \left[\frac{\partial^2}{\partial x^2} + \frac{\partial^2}{\partial y^2} + \frac{\partial^2}{\partial z^2} \right] S(\vec{r},t) +$$

$$v_g \frac{\partial S(\vec{r},t)}{\partial z} \qquad (8.3)$$

where D_a = is the diffusion coefficient in the air,
with other symbols as above.

In the three-dimensional case, the 'diffusion
theory' is represented by different sets of equations
in each of the three regions (soil, crop, and air above
the crop) of the vertical direction. At the boundaries
of these regions equations change qualitatively. To
handle this situation by numerical solution, equations
must be solved separately in the crop region, with the
appropriate boundary condition for soil, and in the air
above the crop region. The assumption that D is
constant within each of these regions is necessary.
Then, solutions can be adjusted to each other by
equating fluxes in the transition layer crop-air by
appropriate boundary conditions.

8.4 THE MULTIPLE-DISPERSAL MECHANISM

Following other phytopathologists and mathematicians
who studied development of plant epidemics, we assumed
that only one mechanism is 'responsible' for disease
spread. But in practice this cannot be so. There is
overwhelming evidence, that two or even three different
mechanisms are involved in disease spread.

All spores start their dispersal within the canopy.
However, some of them can leave the canopy region and
then they are dispersed in the air above the crop. The
dispersal which takes place entirely within the crop
canopy spreads spores over short distances; this
dispersal mechanism will be called the 'short
mechanism'. Spores which left the crop are dispersed
over medium distances in the air above the canopy
region, by the mechanism which will be called the 'long
mechanism', before they are deposited on the crop. If
long-range dispersal should be taken into
consideration, as a third mechanism, the long-distance
spore transport high up in the atmosphere (Zadoks and

Schein, 1979; Pedgley, 1982; Jeger, 1985a) must also be considered. The importance of the multiple-dispersal mechanism was indicated by Vanderplank in his 1975 book (page 137): "Either steep gradients only or shallow gradients only would serve the pathogen badly. ... A mixture of shallow and steep gradients means that the pathogen dispersing along steep gradients could colonize any susceptible plants or fields it found after dispersing along shallow gradients."

Dutch elm disease caused by *Ceratocystis ulmi* offers an example, where the mechanisms involved are not only of different scale, but also of different nature. One mechanism of spread is the infection of healthy plants by root contact with diseased plants; a directional non-random dispersal. Beetles disperse the fungus from diseased to healthy trees over medium distances. A third dispersal mechanism is said to be `responsible' for long-range transport. Beetles are displaced by cars and released at far-away petrol stations (F. Holmes, pers. comm.); a directional partially random dispersal which cannot be described as a diffusion process.

Of course, not all of these dispersal processes are diffusion processes as described in the previous chapters. Even so, some can be mimicked by diffusion. In the case of the Dutch elm disease, dispersal of beetles can be mimicked by diffusion, as beetles released from a `point source' disperse in a diffusion-like way (Wetzler and Risch, 1984). Whether other dispersal mechanisms can be mimicked by diffusion depends on their nature and/or their pattern of dispersal. PODESS is so flexibe that even those dispersal mechanisms which cannot be described or mimicked by diffusion, can be handled in principle.

The above discussion suggests that dispersal of a disease often is due to a multiple-dispersion mechanism, and that this multi-mechanism can sometimes be described by a `multi-diffusion' process. A 'multi-diffusion' process is handled here as a superposition of two or more diffusion processes with different diffusion parameters and different probabilities of occurence.

A model simulating 'multi-diffusion' can be built easily within the framework of the 'diffusion theory'. Each mechanism to be described needs one diffusion equation (of the (3.46) type) with its own values of the diffusion coefficient and the rate of deposition. Exchange of spores between the dispersion mechanisms can be simulated by an exchange term analogous to the deposition or production terms. If lesions resulting from infection by spores, which were dispersed by different mechanisms, should be treated diffrently by the 'diffusion model', separate generalized 'Vanderplank' equations (of the (3.45) type) will be necessary for each dispersal process. The number of equation systems must be equal to the number of dispersal mechanisms involved, when the 'diffusion theory' is applied to the most general case.

8.5 STOCHASTICITY ADDED TO THE MODEL

Plant disease dispersal is a stochastic process at the individual level, the level of the spore and the lesion. Production, dispersal, and deposition of spores, and infection of plant tissue, are stochastic processes. Mathematically, they should be described in probabilistic terms. Numbers of spores produced per unit of time by a sporulating lesion, ratios of germination, colonization and infection, and duration of latency and infectious periods are all random variables (a.o. Mehta and Zadoks, 1970; Eisensmith et al., 1985). Spores are liberated by various processes (Ingold, 1967), which are strongly affected by the local environment, and thus should not be described deterministically. Liberated spores are moved randomly by local air currents and are deposited at random (Tyldesley, 1967; Chamberlain and Chadwick, 1972; Gregory, 1973; McCartney and Fitt, 1985). One cannot say exactly how many spores will be produced and liberated, where every one of them will be deposited, and whether the infection will be successful.

The 'diffusion theory' is a deterministic theory, though it attempts to describe a process which is inherently stochastic. The deterministic approximation is good enough in most cases, but it cannot handle some of the important phenomena which emerge from stochastic events at low inoculum densities. One of them, often observed in field situations, is the appearance of daughter foci. Vanderplank (1975, page 135) stated: "Dispersal over a shallow gradient starts new foci; dispersal over a steep gradient enlarges them". Therefore, an appropriate simulation model of these phenomena is of great importance to plant pathology.

The deterministic approach allows to describe disease development at high densities of spores and lesions. If the density of spores 'in the air' is not high and/or if the probability of infection is low, only a low number of lesions per unit area will be produced. Then, a Poisson distribution with the appropriate mean should be used to 'decide' the number of lesions initialized in an area element and in a small time interval. When the spore production and the effectiveness are high, the diffusion is low and the deposition is strong, stochastic simulation will lead to a ragged boundary of the focus, though the majority of the newly produced lesions will appear close to their mother lesions. At the opposite end low densities of 'distant' spores combined with low probabilities of infection result in very low probabilities of lesion initialization far away from the center of a focus and, consequently, absence of daughter foci. This phenomenon was described by Vanderplank (1975, page 135) as follows: "...dispersal along a steep gradient is unlikely to start daughter foci widely separated from their parents...". If, in contrast, the diffusion is high and the deposition is weak, the distribution of spores 'in the air' is nearly uniform all over the field. Combined with a low spore production and/or with a low probability of infection, this distribution will lead to a nearly uniform distribution of newly produced lesions. Because of low lesion density all over the field, their distribution function cannot be approximated deterministically. Stochasticity will 'produce' random numbers of lesions per grid point. Therefore, the number of lesions in each small area will be a Poisson random variable with a uniform expected density all over the field.

Empirical knowledge tells that a focal disease, initialized by point inoculation, manifests itself as a single mother focus (around the point of initial inoculation) with a nearly circular boundary, and a small number of daughter foci distributed more or less uniformly around the mother focus (Vanderplank, 1975). Real focal disease development looks like the superposition of the pictures resulting from two processes governed by a 'short mechanism' and a 'long mechanism', respectively. The 'short mechanism' builds-up the mother focus and the 'long mechanism' is 'responsible' for the distant spread of disease. The daughter foci resulting from the 'long mechanism' are enlarged by the 'short mechanism'. Some of the spores produced by the daughter foci are dispersed by the 'long mechanism', but they form a small proportion of the total number of spores dispersed by the 'long mechanism'.

The foregoing intuitive argument indicates that stochasticity of lesion initiation together with a double dispersal mechanism are sufficient to simulate focal disease development including the generation of daughter foci.

8.6 DISCUSSION

The 'diffusion theory' originates from a simplified, mechanistic model of spore dispersal. This model leads to the system of two partial differential equations which is the mathematical formulation of the theory. The numerical solution of the equations leads to some phytopathologically useful results. Some of these can be obtained analytically by application of the Diekmann-Thieme theory (Van den Bosch et al., 1988 a, b, c). The advantage of numerical methods is in the combination of a thorough theoretical basis with computer simulation techniques, leading to a flexible computer simulation method. The theoretical background helps to adapt the simulation method to any particular application, and it may also help to avoid some of the 'dead alleys' in a modellers' work. The additional advantage to phytopathology offered by the 'diffusion theory' is its generality, which allows to build a particular simulation model with only slight modification of the original theory. The point was demonstrated in the Sections 8.3 to 8.5, where three possible simulation extensions to the 'diffusion theory' were indicated.

The last but not the least advantage of the approach advocated here is its implementation by means of PODESS (Appendix A), a flexible programming package, allowing to realize a computer program simulating any particular model with very limited knowledge of FORTRAN. The next chapter shows how to convert the models, described above, to the computer programs which simulate phytopathological reality with a fair degree of accuracy.

9 APPLICATIONS

A number of runs simulating various more or less realistic situations of phytopathological interest were performed to show some of the capabilities of the 'diffusion model'. Many specific models, even rather complex ones, can be handled by the software package PODESS (Appendix A). Some programming in FORTRAN is necessary to 'inform' PODESS about the environmental situation and about the equations to be used.

9.1 RICE RESISTANCE BREEDING TRIAL

A promising area of application of the 'diffusion theory' of focus development is computer analysis of plant breeding trials. Experiments dealing with partial resistance need special care in analysis, because of the so-called cryptic error of field experiments (Van der Plank, 1963). This error resides in the overly high severity levels occurring in resistant test varieties (due to net influx of spores from susceptible test varieties), and the overly low severity levels occurring in susceptible test varieties (due to net outflux of spores from susceptible to resistant test varieties).

9.1.1 The method of Notteghem and Andriatompo

A new design for resistance breeding trials was proposed by Notteghem and Andriatompo (1977) for the selection of rice resistant to *Pyricularia oryzae*. The test varieties (cv. Iguape cateto, Kagoshima hakamuri, Aichi asahi, K2, 63-83, and Rojofotsy 1285) were grown in plots (each plot contained three 2.0 m long rows with 0.1 m distance between rows), separated by similar plots of a highly resistant variety (cv. Daniela). The rows were sown parallel to the prevailing wind direction. Perpendicular to the row direction, at the upwind side, a 4.0 m long and 1.0 m wide band of spreader (a susceptible variety) was planted. The density of sowing was 3 g of seeds per meter for the test varieties and the resistant separating variety, and 5 g of seeds per meter for the spreader. The spreader was inoculated 29 March with about $3 \cdot 10^7$ spores of *Pyricularia oryzae*. The degree of severity was measured for each test variety in five points (0.25, 0.5, 1.0, 1.5, and 2.0 meters distance from the spreader along the rows of the test varieties) at five dates in April (8th, 13th, 16th, 22th, and 26th). Disease severity was expressed in degrees according to the 'Bidaux scale'.

The 'diffusion model' requires more parameters than those given by Notteghem and Andriatompo. Their paper gives only a general characterization of the test varieties, the types of lesions covering them, the spatial distribution of the crop, and the distribution of deposited spores. This information allowed only to estimate the area of a single lesion and two spore 'distribution' parameters D and δ. The daily rate of spore production by a single sporulating lesion, the latency period and the infectious period were estimated on the basis of data by Kato (1974) using the general characterization of the varieties given by Notteghem and Andriatompo. The spore distribution parameters, D and δ, were estimated using the method of Williams (1961). The other parameters (i. e. probability of infection, fraction of spores removed from an epidemic, and the leaf area index) we guestimated according to the susceptibility of the varieties, their spatial distribution and the density of sowing. The values of some parameters are shown in Table 9.1. The wind speed $w = 0.7$ [m/day] (this value is very low, but it is the mean wind effect per day). Differences between D, δ and LAI for spreader and test varieties are due to a relatively dense stand of the spreader (the spreader was sown at higher density than the test varieties and the separating variety).

Table 9.1. Numerical simulation of a resistance breeding trial in rice (Notteghem and Andriatompo, 1977). Entries are parameter values used. AREA - area of a lesion [mm^2], LAI - leaf area index, other symbols are explained in Chapter 3.

Variety	Parameter							
	E	D	δ	R	AREA	LAI	p	i
Aichi asahi	0.0098	0.2	3.1	3000	20	4	7	9
6383	0.00294	0.2	3.1	150	1	4	11	5
Kagos. hakam.	0.00588	0.2	3.1	1050	7	4	9	7
Rojo. 1285	0.0098	0.2	3.1	1800	12	4	7	9
Iguape cateto	0.00588	0.2	3.1	300	2	4	9	7
K 2	0.0098	0.2	3.1	3000	20	4	7	9
Daniela	0.00098	0.2	3.1	0	0.5	4	13	3
Spreader	0.0098	0.16	4.0	3000	20	5	7	9

9.1.2 Results

Assessment of disease severity (% of foliage covered by lesions) was performed 10, 15, 18, 24, and 28 days after the first inoculation of the spreader. Results are presented by Notteghem and Andriatompo (1977) in their Fig. 3. The majority of the resulting experimental data and the predictions for the 'diffusion model' are similar or only slightly different (Table 9.2, Fig. 9.1 and 9.2). However, in a few cases, a two-degree difference is observed. This difference reflects the poor knowledge of the parameter values as well as stochasticity of the real process.

Table 9.2. Numerical simulation of a resistance breeding trial in rice (Notteghem and Andriatompo, 1977). Experimental results of a resistance breeding trial (first entry per cell) and numerical results of the 'diffusion model' (second entry) 28 days after the first inoculation of the spreader. Entries are degrees of the 'Bidaux scale' (Notteghem and Andriatompo, 1977, Table III). 0 = no disease (resistant variety), 5 = highly diseased (susceptible variety).

Variety	Distance from spreader											
	0.00		0.25		0.50		1.00		1.50		2.00	
6383	1	2	1	1	1	1	1	1	0	0	0	0
Iguape c.	3	2	3	2	3	2	2	1	2	1	2	0
Kagoshima	4	3	4	2	4	2	3	2	3	1	2	1
Aichi asahi	4	5	4	4	3	4	3	3	2	2	2	2
K2	4	5	4	5	5	4	4	3	3	2	2	2
Rojof. 1285	5	4	5	4	5	3	4	2	4	2	3	1

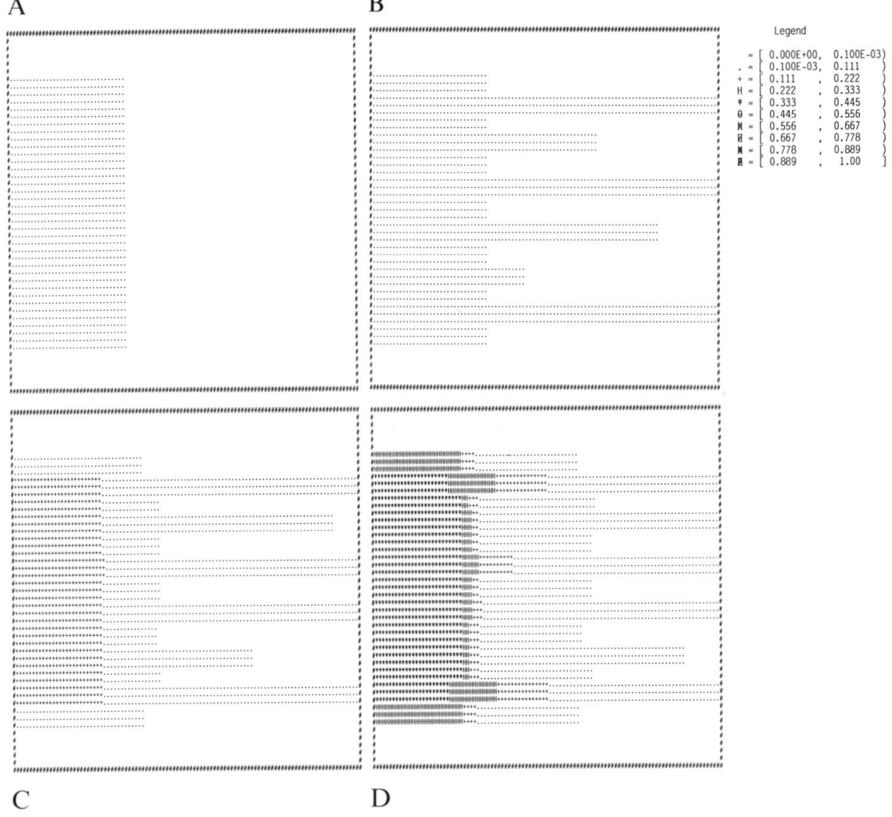

Fig 9.1. Numerical analysis of a resistance breeding trial in rice (Notteghem and Andriatompo, 1977). Development of disease in a breeder's trial field during run 1 for 0, 18, 28 and 36 days after initial inoculation. Rows of the test varieties are along the X-axis, the spreader is placed along the Y-axis on the left-hand side. X-axis is a distance from 0 to 3 m, and Y-axis is a distance from 0 to 4.8 m. Intensities of printed points reflect the fractions of plant surface covered by lesions. **A**, T = 0. **B**, T = 18 days after initial inoculation. **C**, T = 28 days after initial inoculation. **D**, T = 36 days after initial inoculation.

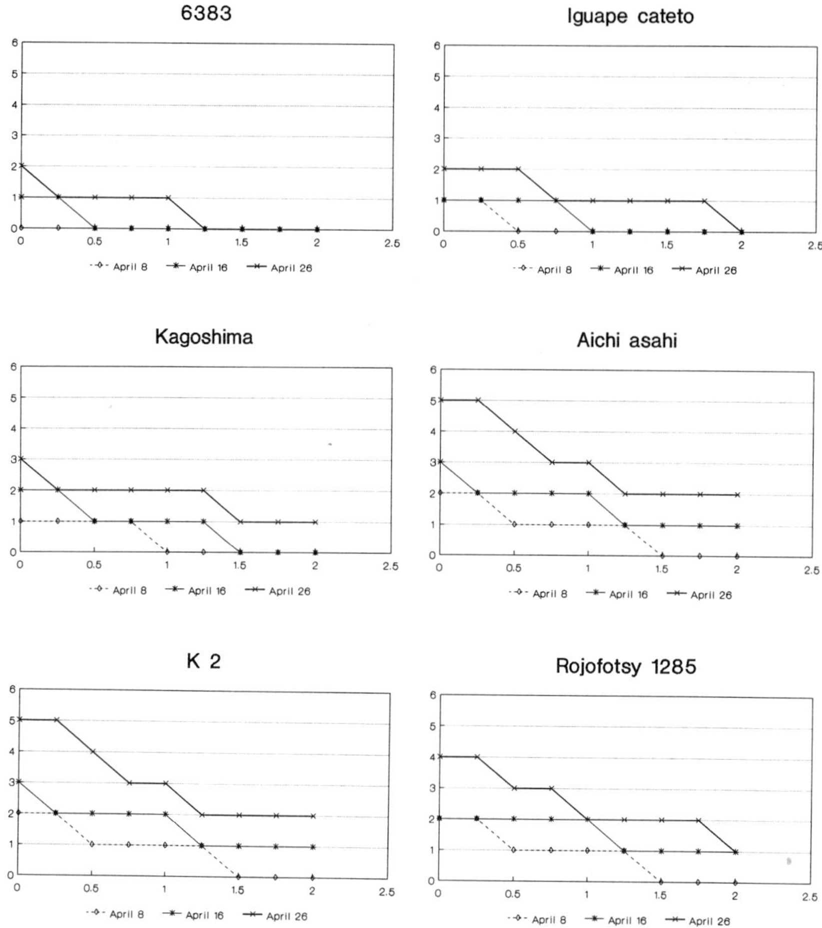

Fig 9.2. Numerical analysis of a resistance breeding trial in rice (Notteghem and Andriatompo, 1977). Simulated severity of disease on test varieties during run 1. Curves are printed for: 8, 16, and 26 april. The X-axis represents the length (in meters) of the test variety row, the Y-axis represents degrees of disease severity. Compare Fig. 3 of Notteghem and Andriatompo (1977).

An attempt was made to assess the cryptic error,
sensu Van der Plank (1963), of the breeding trial. The
following runs of the program were performed:
1. Original trial.
2. A susceptible variety (Aichi asahi) at every
   position of the test varieties.
3. A susceptible variety (Aichi asahi) at every
   position of the test varieties and at the positions
   of the separating variety.
4. A susceptible variety (Aichi asahi) covers the whole
   experimental field.
5. A resistant variety (6383) at every position of the
   test varieties.
6. A resistant variety (6383) at every position of the
   test varieties and at the positions of the
   separating variety.
7. A resistant variety (6383) covers the whole
   experimental field.

Table 9.3 shows the results for runs 1, 2, 3, and 4.
Relative progress of the disease severity can be judged
by comparison of the runs. Replacement of all the test
varieties by the susceptible one increases the net
influx from the variety neighbourhood (run 2).
Replacement of the separating variety has a similar
effect (run 3). Replacement of the spreader by a
susceptible variety increases its severity level at the
observation position, because the original spreader was
sown more densely than the test varieties, so that
fewer spores than after replacement could diffuse to
the part of the field originally covered by the test
varieties (run 4).

Analogous runs for a resistant variety (Table 9.4,
runs 1, 5, 6, and 7), give a very different picture.
Replacement of the test varieties by a resistant one
(run 5) and, additionally, the separating variety by a
resistant one (run 6) has no observable effect. The

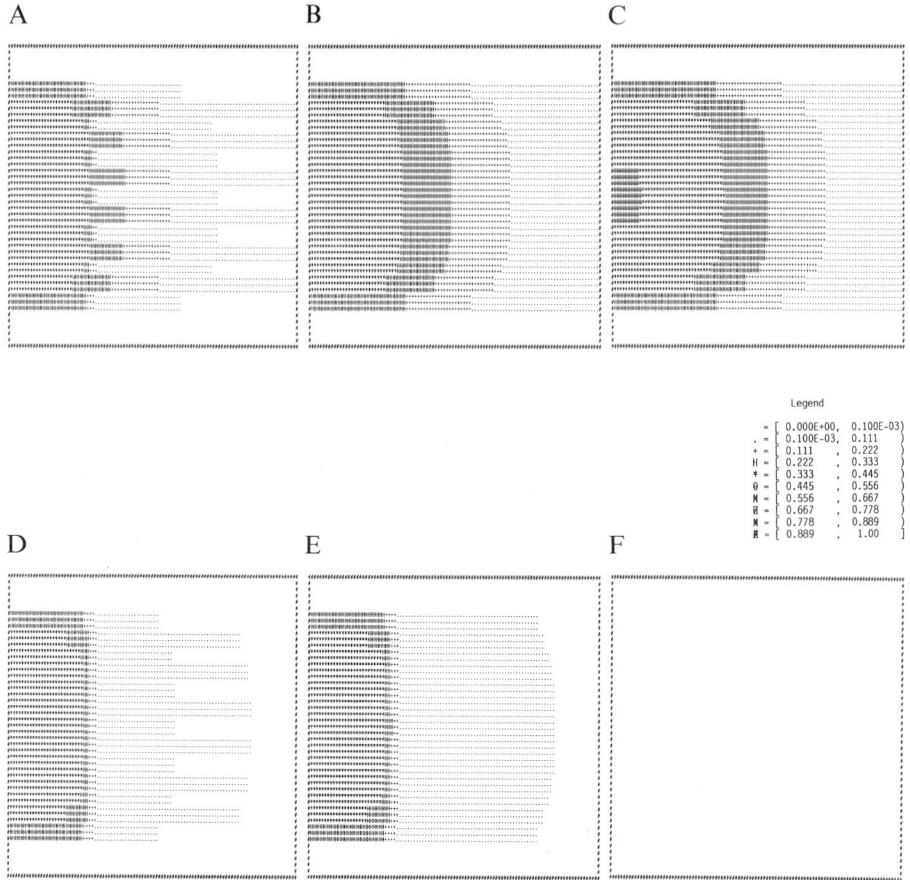

Fig 9.3. Numerical analysis of a resistance breeding trial in rice (Notteghem and Andriatompo, 1977). Simulated disease severity on a breeder's trial field for runs 2, 3, 4, 5, 6, and 7, respectively, 36 days after initial inoculation. Legend as for Fig. 9.1, but for conditions described in the text (specified with the run number). A, results of run 2. B, results of run 3. C, results of run 4. D, results of run 5. E, results of run 6. F, results of run 7.

result of run 5 is equivalent to a low effect of the susceptible test varieties on the resistant ones. The result of run 6, in which a resistant separating variety was replaced by another resistant variety, adds nothing to the results of run 5. In these cases the words `low effect' instead of `no effect' are used, because the scale values give only a rough picture of the real severities, though maybe good enough from a breeder's point of view. The only situation which influences the results is replacement of the spreader by a resistant variety (run 7). In this case disease does not develop. Severities obtained by the compared runs at T = 36 days after inoculation are shown on Fig. 9.3.

Table 9.3. Numerical simulation of a resistance breeding trial in rice (Notteghem and Andriatompo, 1977). Results of the runs for the susceptible variety (see text), Aichi asahi, 28 days after initial inoculation. Replacement of the trial elements by the susceptible variety leads to higher disease severities. Entries are degrees of the `Bidaux scale' (Notteghem and Andriatompo, 1977, Table III). 0 = no disease (resistant variety), 5 = highly diseased (susceptible variety).

Distance from spreader	Run 1	Run 2	Run 3	Run 4
0.00	5	5	5	5
0.25	4	5	5	5
0.50	4	4	4	5
1.00	3	3	4	4
1.50	2	2	2	2
2.00	2	2	2	2

Table 9.4. Numerical simulation of a resistance breeding trial in rice (Notteghem and Andriatompo, 1977). Results of the runs for the resistant variety (see text), 6383, 28 days after initial inoculation. Entries are degrees of the `Bidaux scale' (Notteghem and Andriatompo, 1977, Table III). 0 = no disease (resistant variety), 5 = highly diseased (susceptible variety).

Distance from spreader	Run 1	Run 5	Run 6	Run 7
0.00	2	2	2	0
0.25	1	1	1	0
0.50	1	1	1	0
1.00	1	1	1	0
1.50	0	0	0	0
2.00	0	0	0	0

9.1.3 Discussion

Comparison of the results of the numerical simulation with the experimental results of Notteghem and Andriatompo indicates good qualitative and quantitative consistency. The differences are probably due to the simplifying assumptions about the parameters and to not fully correct guesses of the parameter values. The consistency suggests that the estimated parameter values are close to their true values.

The results shown in Table 9.2 and formulated in the preceding section warrant a fundamental analysis of resistance breeding trials and an estimation of their cryptic error. The qualitative behaviour of the solutions is consistent with the opinion about the

cryptic error forwarded by Van der Plank (1963).
Especially evident is the cryptic error for resistant
variety 6383, comparing the cases with presence and
absence of the spreader. Without spreader, disease
severity level is nil (degree 0), whereas with spreader
it is considerable (degrees 1 and 2).

It should be explained why differences of 2 degrees
between the experimental and the numerical results are
treated as negligible in Table 9.2, whereas for the
results shown in Tables 9.3 and 9.4 a difference of one
degree is considered important. In the first case,
uncertainty about the parameter values, stochasticity
of the real process, and the possible influence of
factors not described by the 'diffusion theory', can
have so considerable an effect, that even a difference
of two degrees between the experimental and the
numerical results seems acceptable. In the second case,
only computer results are compared. These runs were
performed by solving the same equations with the same
parameter values in all runs. The only differences were
those in trial design. So, the differences observed in
Tables 9.3 and 9.4 are treated as reflecting the effect
of differences in trial design.

9.2 DOUBLE DISPERSAL MECHANISM –
INTENSIFICATION AND EXTENSIFICATION OF AN EPIDEMIC

A model simulating a double dispersal mechanism,
i.e. a combination of the 'short' and the 'long'
mechanisms, was realized according to the ideas
explained in Section 8.4. The two dispersal mechanisms
are common in air-borne plant diseases. The 'short
mechanism' is characterized by:
1. a low value of the diffusion coefficient, D,
2. a high value of the rate of spore deposition, δ, and
3. a high fraction of the spores involved.
This mechanism models the within-crop dispersal with:

a. a low value of the wind speed and small eddies,
b. spores have a high chance to be deposited, as they
   are close to crop surfaces,
c. most of spores are dispersed within the crop canopy.
The `long mechanism' is characterized by:
1. a high value of the diffusion coefficient, D,
2. a low value of the rate of spore deposition, δ, and
3. a low fraction of spores involved (those which were
   not dispersed by the `short mechanism').
This mechanism models the above-crop dispersal with:
a. a high value of the wind speed and large eddies,
b. spores can travel a long distance before they return
   to the crop layer where they are deposited,
c. only those spores which could temporarily `escape'
   from the crop are dispersed by this mechanism.

9.2.1 Crop pattern and parameters

The simulated field was a square of 100 x 100 m,
represented by a grid of 31 x 31 points. The field was
covered uniformly by a susceptible crop.

As the lesions `behaved' identically irrespective of
the dispersal mechanism of the spores which generated
them, two diffusion equations (one per mechanism) and
one `generalized Vanderplank equation' were used. Only
one set of parameters characterizing lesions and two
sets of parameters characterizing spores were used. The
parameters describing lesions were:
1. R - reproductivity = 4 [daughter lesions per
                          sporulating mother lesion per
                          day],
2. p - latency period = 3 [days],
3. i - infectious period = 5 [days],
4. A - area of a single lesion = 10 [mm^2].
The `spore parameters' for the `short mechanism' were:
1. D_1 - diffusion coefficient = 1 [m^2/day],
2. δ_1 - rate of spore deposition = 10 [1/day],

3. E_1 - effectiveness = 1.

The `spore parameters' for the `long mechanism' were:

1. D_2 - diffusion coefficient = 150 [m^2/day],

2. δ_2 - rate of spore deposition = 0.3 [1/day],

3. E_2 - effectiveness = 1.

The leaf area index LAI = 5.

Thirteen runs were performed for the following proportions of spores dispersed by the `short mechanism', $F$ = 0, 0.01, 0.1, 0.2, 0.3, 0.4, 0.5, 0.6, 0.7, 0.8, 0.9, 0.99 and 1. The remaining spores were dispersed by the `long mechanism'. For each run, the crop was `inoculated' at the centre of the field by a single successful spore at time T = 0.

9.2.2 Results

Results were obtained by running PODESS on the VAX 785 computer of the Wageningen Agricultural University. Each run lasted 50 `simulation days'. Every 10 `simulation days', the values of the two spore distribution functions and the lesion distribution function were printed and plotted.

A comparison of the lesion density distributions produced by these runs shows the overriding influence of the partitioning of spores over the two dispersal mechanisms. For a closer examination, three functional dependencies of spore partitioning were examined: (1) the velocity of focus expansion, (2) the lesion densities at points with different distances from the focal centre, and (3) the total number of lesions present in the field.

The velocity of focus expansion shows a fast decrease with increasing proportions of spores dispersed by the `short mechanism' in the range of F = 0 to F = 0.2, a low effect of partitioning of spores in the range from F = 0.2 to F = 0.99, and a rapid drop with F changing from 0.99 to 1 (Fig 9.4).

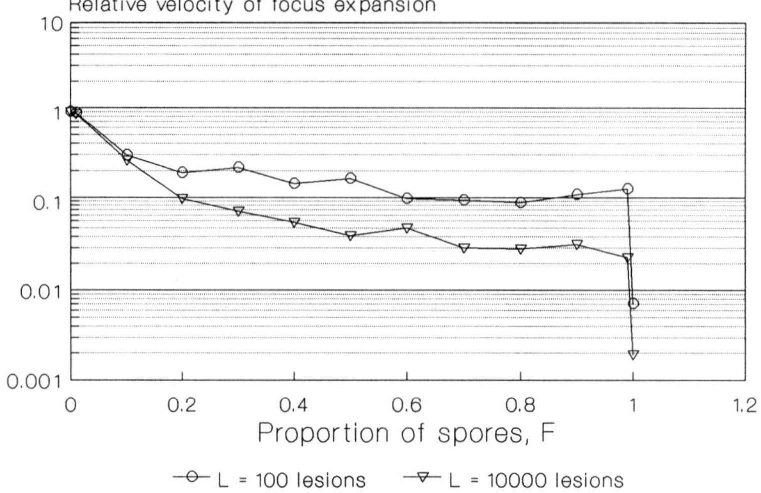

Fig. 9.4. Focus expansion with a double dispersal mechanism in a deterministic situation. The x-axis represents the proportion of spores dispersed by the `short mechanism', F. The y-axis represents the scaled velocity of focus expansion (the highest velocity equals 1) on a logarithmic scale. The velocities were determined at two 'levels' of the front of the epidemic (see Section 3.4.2): L_o = 100 lesions and L_o = 10000 lesions.

Lesion density at a certain time varies strongly with the distance of the measurement point from the origin. Fig. 9.5. shows results obtained for the `simulation time' T = 40. Lesion density at the centre and in its close vicinity grows with F in an S-shaped manner. Far from the centre, the lesion density grows fast with increasing F, reaches its maximum at F = 0.8 and then decreases rapidly when F increases still further.

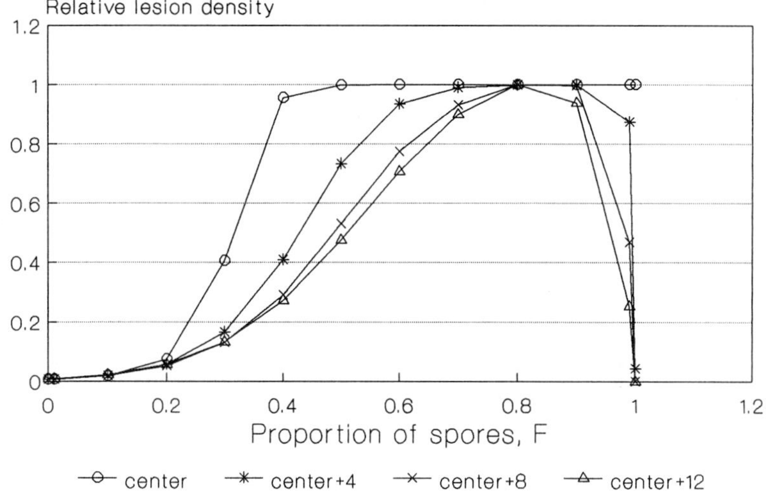

Fig. 9.5. Focus expansion with a double dispersal mechanism in a deterministic situation. The x-axis represents the proportion of spores dispersed by the `short mechanism', *F*. The y-axis represents the relative lesion density (the highest density per distance equals 1) at the centre and at three points with distances from the centre: 4, 8 and 12 units of distance between grid points (or 13.2, 26.4, and 39.6 m).

The total number of lesions in the focus for successive values of time (Fig. 9.6) increases exponentially with *F* in the early stages of epidemic growth. Later, an interesting phenomenon is observed; with growing value of *F*, the total number reaches a maximum (for *F* = 0.9 at T = 30 and for *F* = 0.8 at T = 40) and then decreases. The curve of the total number of lesions for T = 50 reaches its maximum value at *F* = 0.2, stays there until *F* = 0.99, and suddenly drops down with *F* passing from 0.99 to 1.0.

153

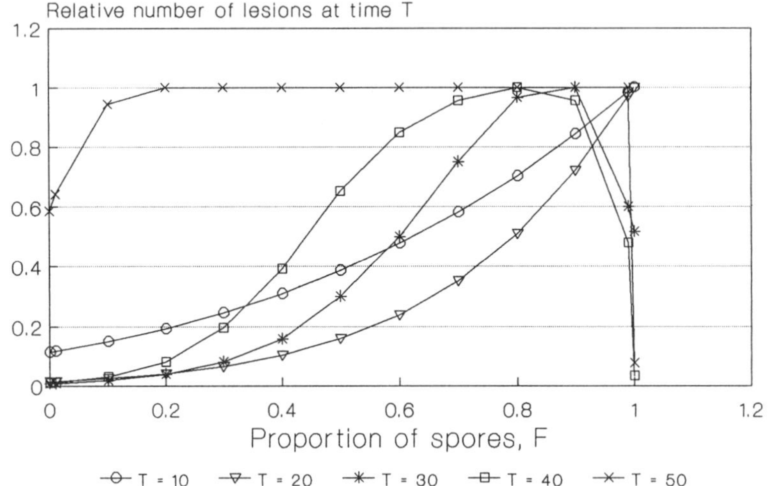

Fig. 9.6. Focus expansion with a double dispersal mechanism in a deterministic situation. Relative total number of lesions per field at five times: T = 10, 20, 30, 40 and 50. The x-axis represents the proportion of spores dispersed by the `short mechanism', F. The y-axis represents the relative total number of lesions (of the whole field) expressed in scaled values (the highest value equals 1).

After an initial phase (up to T = 20) where the total number of lesions increases exponentially with F, its response to F is visualized by a curve having a relative sharp peak at $F = 0.9$ on T = 30. With the increase of time, the maximum broadens, and its peak value is attained at lower values of F, until at T = 50 the curve loses its peak and shows a broad maximum plateau from $F = 0.2$ to $F = 0.99$.

154

9.2.3 Discussion

The phenomena shown in Fig. 9.4 to 9.6 have a common explanation. The lesion density at the centre grows in an S-shaped manner with the proportion of spores dispersed by the `short mechanism', $F$, (Fig. 9.5); this region of the field is influenced almost exclusively by the `short mechanism'. When the lesion density reaches the maximum of sites available, growth is limited by exhaustion of vacant sites. At the points more distant from the centre, the influence of the `short mechanism' is weaker, as shown by the decrease of the lesion density with F above the maximum value, $F = 0.8$ at T = 40. Accordingly, the focus expansion velocity, which measures the spatial progress of the focus boundary at relatively low lesion densities, decreases fast with F changing from 0 to 0.2 and drops steeply when F changes from 0.99 to 1. This drop is the result of excluding the `long mechanism' from spore dispersal. Then, the conquest of space by a disease becomes much slower, as also shown by the functional dependence of the total number of lesions on F (Fig. 9.6).

In an infinite field, the total number of lesions present in the field, $\mathfrak{L}$, decreases with increasing F, as this allows a continual exploitation of fresh sites which have not been exhausted yet. Therefore, the best strategy would be $F = 0$ (i.e. only the `long mechanism'). In a finite field, with only one dispersal mechanism, $\mathfrak{L}$ decreases with growing value of $\mathfrak{U}$ (the ratio of field size to the width of the contact distribution), as long as vacant sites are not exhausted. For that situation the best strategy is $F = 1$ (i.e. only the `short mechanism'). Apparently the two mechanisms, site exhaustion and spore loss to the hostile environment outside the field, yield opposite results.

Real fields are finite and have a finite density of

sites. After an initial phase, exhaustion of available sites in the centre becomes limiting and spatial effects begin to dominate. As a result of the two effects, (1) loss of spores blown outside a field and (2) exhaustion of available sites at the focus centre, neither $F = 0$ nor $F = 1$ are the best strategy, but a value of F in between. Where the total number of lesions grows quadratically with the radius of a focus and linearly with the lesion density per point, the total number of lesions grows exponentially with F in the early stages, when the lesion density at the centre depends exponentially on F. Later, with the exhaustion of available sites at the centre, spatial expansion of the focus becomes more important to augment $\mathfrak{L}$, a large area covered by the focus having more influence on total number of lesions than the centre and its close vicinity. The curves of the total number of lesions, $\mathfrak{L}$, for T = 30 and T = 40 show maxima at certain values of F, due to the interaction of the two dispersal mechanisms. The maximum for F = 0.8 at T = 40 corresponds to the maxima shown by most curves of lesion density at T = 40 (Fig. 9.5). The last curve for total number of lesions, at T = 50, shows a wide region of F where it keeps its maximum value, a result of the exhaustion of available sites all over the field. The sudden drop of the total number of lesions, shown by this curve when passing from F = 0.99 to F = 1, proves once more that dispersal by the 'short mechanism' alone is not the most efficient way of focus expansion.

The 'short mechanism' alone causes 'intensification' (Zadoks and Kampmeijer, 1977) of a disease within a relatively small area. Severity increases fast near to the infected point, but spatial development of disease is slow. Quite oposite are the results of the 'long mechanism'. It spreads disease over a large area, with hardly any intensification; this is 'extensification' (Zadoks and Kampmeijer, 1977). A real epidemic of an

air-borne plant disease is spread by both the `short'
and `long' mechanisms, which interact. New areas are
infected by spores dispersed by the `long mechanism'
and then the disease intensifies there by the `short
mechanism'. Due to the interaction of the two
mechanisms on a field of finite size with a finite
density of sites available for infection, epidemic
growth is much faster than summation of the results of
the two mechanisms might suggest. The interaction
effect is clearly illustrated by the maxima, usually at
about F = 0.8, appearing in the Figures 9.5 and 9.6.

The above examples relate to a dual dispersal
mechanism. A triple dispersal mechanism can also be
programmed. Multiple dispersal mechanisms, to use a
general term, are well known in phytopathology but they
have hitherto been neglected as phenomena to be studied
per se. They considerably speed up disease progress, by
joint extensification and intensification. One possible
reason for the neglect of multiple dispersal is the
very small fraction of spores needed for an effective
`long mechanism', a fraction usually much below the
threshold value which is one daughter lesion per mother
lesion (the `threshold theorem', Van der Plank, 1963).

It seems plausible that, by taking the multiple
dispersal mechanism into consideration, the degree of
realism of models simulating plant disease development
will increase considerably. The next section discusses
a model which, by applying the multiple dispersal
mechanism in combination with randomization of lesion
initialization, allows to simulate a real life
phenomenon - the appearance of daughter foci.

9.3 DOUBLE DISPERSAL MECHANISM + STOCHASTICITY =
DAUGHTER FOCI

The spores dispersed by the `long mechanism' are
spread over a large area in comparison to spores

dispersed by the `short mechanism'. Therefore, the spore density and so the newly initialized lesion density is low, even with a high proportion of spores dispersed by the `long mechanism'. Under these conditions the deterministic approach of the `diffusion theory' can no longer be applied. The number of lesions initiated at a certain point by `long mechanism' spores is a stochastic variable with a Poisson distribution.

A stochastic approach is also necessary when spores dispersed by the `long mechanism' remain air-borne for a longer period, during which they are exposed to environmental conditions which decrease their infectivity (Jeger, 1985a). Therefore, the probability of infection, $I$, for spores dispersed by the `long mechanism' will be lower than that for spores dispersed by the `short mechanism'. As $I$ affects the fraction of spores which will initiate lesions, the `long mechanism' has a lower effectiveness, E (sensu Zadoks and Schein, 1979). Thus, even relatively high numbers of spores dispersed by the `long mechanism' can result in low numbers of newly-produced lesions.

9.3.1 Crop pattern and parameters

A simulation model was designed with a square field of 100 x 100 m, represented by a grid of 31 x 31 points. The field was uniformly covered by a susceptible crop. The model describing superposition of the two dispersal mechanisms contained two systems of `diffusion theory' equations, analogous to the system (3.45), (3.46) (one system per mechanism). The spore dispersal for each mechanism was described by a diffusion equation (of (3.46) type). The stochasticity of lesion initialization by `long mechanism' spores was simulated by means of a random number generator which `decided' how many lesions were produced. The random numbers generated had a Poisson distribution with the

expected number of newly produced lesions as its
parameter. A second random number generator, with
uniform distribution of generated numbers, `decided'
the exact positions of newly initiated lesions. The
procedure was used once per simulation day, as long as
the number of newly produced lesions was low. If it was
high, the deterministic approach was applied. As an
upper limit of the stochastic approach, the number of
10000 newly initiated lesions per day was used. This is
an arbitrary choice. Rather than the total number the
local density of newly initialized lesions should be
examined to decide whether or not to use the stochastic
approach. As spores are distributed nearly uniformly
over the field by the `long mechanism', both ways are
applicable, and the total number criterion is faster.
The density of lesions initiated by `short mechanism'
spores was treated as a deterministic function. The
number and positions of the newly produced lesions thus
generated, in other words the newly generated lesion
density function, were inserted into the two
generalized Vanderplank equations (of (3.45) type), one
per mechanism, with a common term correcting for
removals

$$(1-(L_1+L_2)/L_{max})$$

L_1 is the density of lesions initialized by the `short
mechanism', L_2 is the density of lesions initialized by
the `long mechanism' and L_{max} is the density of sites.
These equations describe changes in the rate of lesion
production. Since lesions `behave' identically
irrespective of the dispersal mechanism of the spores
which generated them, only one set of parameters
characterizing lesions and two sets of parameters
characterizing spores were used. The `lesion
parameters' were (see Section 9.2.1):
1. R - reproductivity = 4 [daughter lesions per

sporulating mother lesion per day],

2. p - latency period = 3 [days],
3. i - infectious period = 5 [days],
4. A - area of a single lesion = 10 [mm^2].

The 'spore parameters' for the 'short mechanism' were:

1. D_1 - diffusion coefficient = 1 [m^2/day],
2. δ_1 - rate of spore deposition = 10 [1/day],
3. E_1 - effectiveness = 1.

The 'spore parameters' for the 'long mechanism' were:

1. D_2 - diffusion coefficient = 150 [m^2/day],
2. δ_2 - rate of spore deposition = 0.3 [1/day],
3. E_2 - effectiveness = 0.2.

The leaf area index LAI = 5.

Eight runs were performed for the following proportions of spores dispersed by the 'short mechanism', F = 0, 0.2, 0.4, 0.6, 0.8, 0.9, 0.99 and 1. For each run, the crop was 'inoculated' at the centre of the field by a single successful spore at time T = 0.

9.3.2 Results

Results were obtained by eight runs of PODESS. Each run lasted 50 simulation days. Every tenth day, the values of the lesion distribution function were printed and plotted.

The results showed qualitative differences between the lesion distribution functions obtained by runs with different values of F because of the appearance of daughter foci, scattered randomly around a relatively big mother focus. The pictures of the lesion density distribution at time T = 40 for different values of F indicate that daughter foci appear between F = 0.4 and F = 0.99 (Fig. 9.7). The most 'realistic' effect seems to be obtained at values of F between 0.8 and 0.99.

The influence of F on focus development is shown by three characteristics of spore partitioning: (1) the

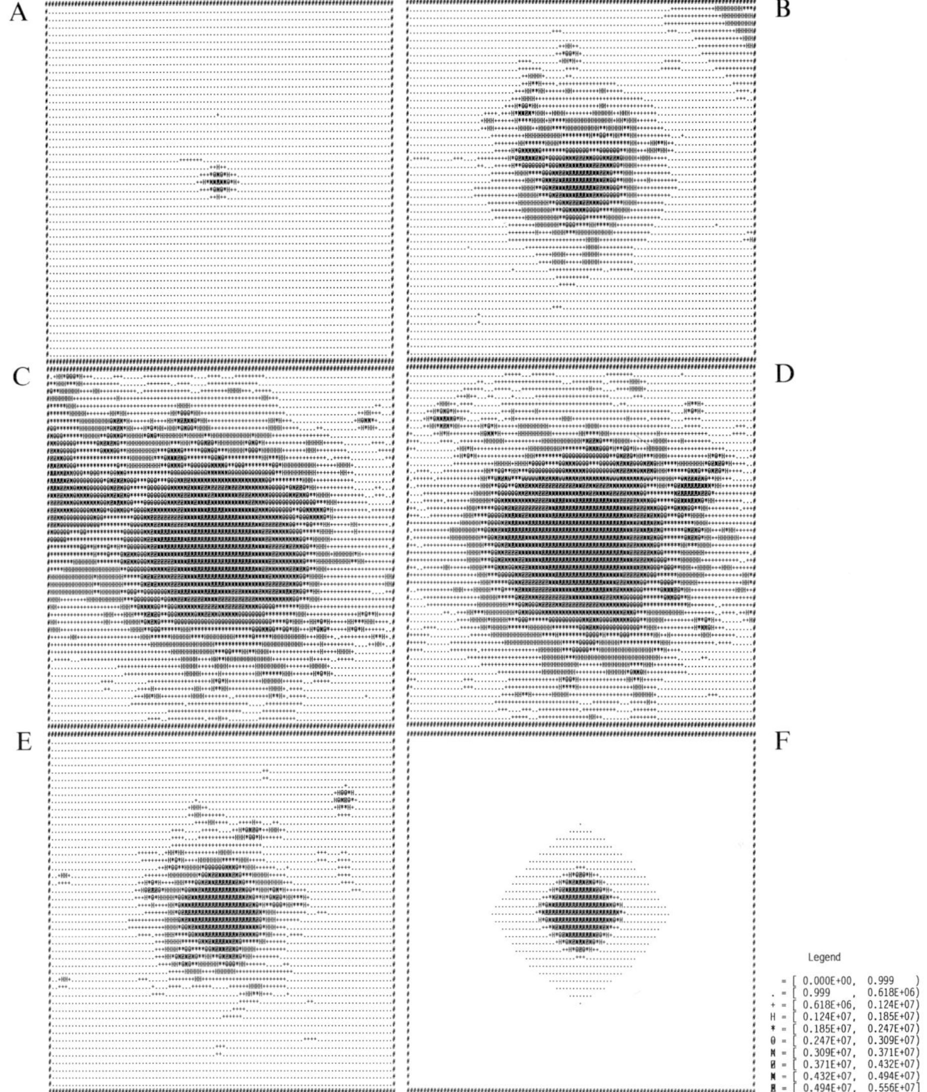

Fig. 9.7. Focus expansion with a double dispersal mechanism in a stochastic situation. The lesion density distribution at time T = 40 for different values of F. X- and Y-axes are distances from 0 to 100 m, intensity of printed points reflects the number of lesions on the host surface. **A**, $F = 0.4$. **B**, $F = 0.6$. **C**, $F = 0.8$. **D**, $F = 0.9$. **E**, $F = 0.99$. **F**, $F = 1$.

frequency distribution of lesion densities at time T =
40, (2) the total number of lesions in three regions of
the field at time T = 40, and (3) the total number of
lesions present in the field at five different times.

The range of possible lesion densities (between 0
and the maximum lesion density) was divided into eight
classes. A frequency distribution over these eight
classes was based on all 31 x 31 grid points
representing the field. The frequency distribution was
scaled by giving the frequency of the best filled class
the relative frequency value 1. The scaled values were

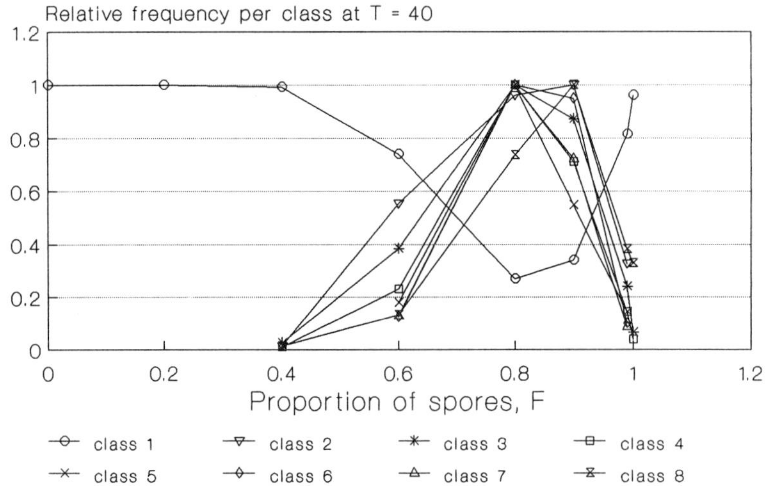

Fig. 9.8. Focus expansion with a double dispersal
mechanism in a stochastic situation. Relative
frequencies of lesions per frequency class plotted as
functions of the proportion of spores dispersed by the
`short mechanism', *F*. Class 1 represents the grid
points with low number of lesions, class 8 those with
high numbers of lesions. *F* is plotted along the x-axis.
The y-axis represents the relative frequency of lesion
densities per class at time T = 40 (the highest
frequency per class equals 1).

162

plotted against F (Fig 9.8). The maximum of the low density class (class 1) appears for low values of F and for $F = 1$. The other classes have their maxima around $F = 0.8$ or 0.9. The result indicates that a partitioning of the available spores over the `short mechanism' and the `long mechanism' in a proportion of about 85 : 15 is optimal for the production of highly infected points, under the model conditions specified.

The dependence of the total number of lesions in the field on F, for different values of time, T, is shown in Fig 9.9, which is a stochastic counterpart of Fig. 9.6. In the early stages of epidemic growth the total number of lesions per field for successive values of time increases exponentially with F. Later, at T = 40,

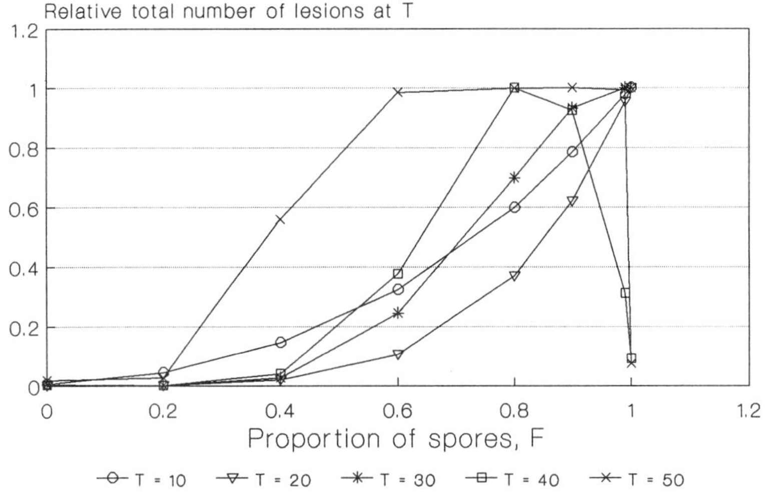

Fig. 9.9. Focus expansion with a double dispersal mechanism in a stochastic situation. Relative total number of lesions per field at five times T = 10, 20, 30, 40, and 50. See Fig. 9.6. The x-axis represents the proportion of spores dispersed by the `short mechanism', F. The y-axis represents the relative total number of lesions (of the whole field) at time T. Compare Fig. 9.6.

a maximum is attained at $F = 0.8$, and at T = 50 the
maximum becomes a plateau from $F = 0.6$ to 0.99. The
stochastic situation of Fig. 9.9 resembles the
deterministic situation of Fig. 9.6, though the peaks
or plateaux of the maxima are narrower.

To examine the frequency distribution of lesion
densities over the field, three curves for the total
number of lesions were plotted versus F at time T = 40
(Fig. 9.10): (1) for the whole field, (2) for a
`central region', and (3) for the `peripheral region'.

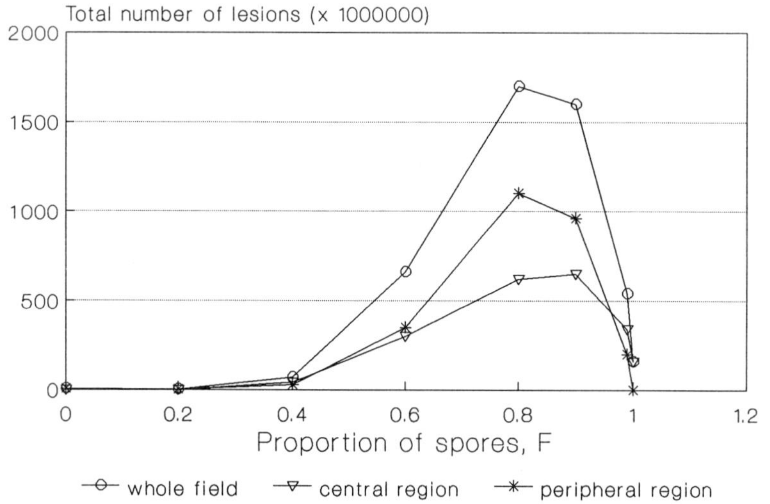

Fig. 9.10. Focus expansion with a double dispersal
mechanism in a stochastic situation. Total number of
lesions for the whole field, for a `central region' and
for a `peripheral region' at T = 40, as functions of F.
The proportion of spores dispersed by the `short
mechanism', F, is plotted along the x-axis. The total
number of lesions is plotted along the y-axis. Compare
Fig. 9.8.

The `central region' was determined by the size of the
mother focus for $F = 1$ at T = 40, the `peripheral
region' stands for the rest of the field. The total
number of lesions per field and the total number of
lesions of the `peripheral region' reach their maximum
at $F = 0.8$. The total number of lesions of the `central
region' has its maximum at $F = 0.9$. At the right hand
side of the maximum, all three curves show a fast
decrease. For $F = 1$, the number of lesions in the
`peripheral region' approaches 0. In other words, focus
formation is limited to a single focus without daughter
foci, the outcome of the purely deterministic
situation.

9.3.3 Discussion

A comparison of the lesion density distributions
produced by the runs in a stochastic situation confirms
the results described in Section 9.2 about the effect
of interaction between the `short' and the `long'
dispersal mechanisms. The explanation of the maximum
effects for certain values of F is identical to the one
presented in Section 9.2.

Interaction between the `short' and the `long'
dispersal mechanisms together with randomization of the
lesion initiation by spores dispersed by the `long
mechanism' allows to simulate not only the phenomena of
intensification and extensification, but also the
phenomenon of daughter foci. The latter phenomenon is
observed only for certain values of spore partitioning
over the two dispersal mechanisms; daughter foci become
visible for values of F above 0.4 and they disappear
again above $F = 0.99$.
The comparison of deterministic and stochastic
situations, as simulated above, is interesting. In the

deterministic situation, the effectiveness E_2 = 1 for spores dispersed by the 'long mechanism'. In the stochastic situation, a lower value of effectiveness E_2 = 0.2 results in higher values of F needed to reach the same level of the total number of lesions per field (see Section 9.2). The shift in the value of F is easy to explain, as the effectiveness influences the proportion of lesions initialized by deposited spores. The common characteristic of both situations is the maximum of the curves for (relative) total number of lesions at F = 0.8 in Fig. 9.6 and 9.9. The importance of the maximum at about F = 0.8 is also indicated by the curves of relative lesion density at points with different distances from the initial inoculation point (Fig. 9.5) and the relative frequency curves (Fig. 9.8).

This comparison of the deterministic and the stochastic versions of the double dispersal mechanism suggests that a value of F between 0.8 and 0.99 has a great epidemiological importance. The optimum value of F depends on the parameters chosen to run the 'diffusion model'. The parameters used here result from a compromise between 'realism' of the model and speed of calculation. In real life, F will be a variable and its value may sometimes be outside the indicated range of 0.8 to 0.99. Fairly realistic model calculations for *Puccinia striiformis* (Rijsdijk and Rappoldt, 1980) arrived at values of about 0.93 for a field with an area of 10000 m^2.

Taking into account the finite density of sites available for infection and the finite field size, a pathogen which disperses its spores by a double dispersal mechanism is much more efficient in the conquest of space. Actually, a pathogen of the 'wind-borne, foliar' type cannot survive without a double dispersal mechanism; it is an obligatory dispersal strategy.

9.4 LEAF RUST ON WHEAT - THE THREE-DIMENSIONAL CASE

In most cases, the two-dimensional representation of
a field is good enough for the analysis of focus
development. If a more detailed picture is necessary,
the influence of the third dimension, crop height, must
be considered. The three-dimensional version of the
'diffusion theory' is adequate for these situations.
Generally speaking, the approach discussed in Section
8.3 should be applied. As the model uses different
equations for the crop region and the air above crop, a
special time consuming procedure is needed to run the
model. If the field size is relatively small, the
above-crop spore dispersal can be neglected and only
the within-crop spore dispersal is to be simulated.
Such a simplification allows to simulate vertical focus
development on developing leaf layers. Wind effect on
focus development can also be included.

As an example of the three-dimensional approach, the
'diffusion model' for the experiment discussed in
Section 7.4 was built. To examine additional effects,
developing leaf layers and prevailind wind direction
were incorporated into the three-dimensional version of
the 'diffusion model'.

9.4.1 The simulation technique

Five vertical layers were distinguished: (1) the
soil, (2) three leaf layers, and (3) the upper crop
boundary. The properties of these layers were:
1. soil - spores are deposited, but they do not
   initialize lesions,
2. leaf layers - spores diffuse within the crop region
   and when deposited on leaves can initialize lesions,
3. upper crop boundary - spores diffuse into that layer
   to simulate their escape to the air above a crop
   region, where they are lost for the simulated

epidemic as they are transported outside the simulated field.

The properties of these three regions were simulated by:

a. the absorbing boundary conditions of the three-dimensional version of the system of equations according to the 'diffusion theory', simulating strong absorption of spores,

b. the three-dimensional version of the system of equations according to the 'diffusion theory', simulating spore dispersion and deposition, lesion initialization, and spore production by sporulating lesions,

c. the three-dimensional version of the system of equations according to the 'diffusion theory', with the value of LAI = 0 simulating spore dispersal, but without deposition, lesion initialization and spore production.

9.4.2 Parameters

As no parameters required by the 'diffusion theory' were measured, parameter values similar to those in Section 7.5 were used. Focus development was simulated during 93 days with the following parameter values:

1. D – diffusion coefficient = 0.02 [m^2/day],
2. δ – maximum rate of spore deposition = 2 [1/day],
3. E – effectiveness = 1,
4. R – productivity = 3.5 [daughter lesions per mother lesion per day],
5. p – latency period = 18 [days],
6. i – infectious period = 16 [days],
7. AREA – area of a single lesion = 10 [mm^2].
8. w – the velocity of wind is given in Table 7.1; for the days between the indicated ones, the wind velocity was calculated by linear interpolation from the two nearest dates,

Table 9.5. The three-dimensional 'diffusion model' with variable leaf layers. The leaf area index, LAI, varies with leaf layer and time. The values are effective LAI available for infection (the experimental data were measured in the Peterson B-scale for which 100% severity equals 37% of infected leaf area; see Zadoks and Schein, 1979).

Time	LAI		
	Leaf layer 1	Leaf layer 2	Leaf layer 3
0	0.5	0.1	0.
10	0.5	0.15	0.
20	0.5	0.2	0.
30	0.5	0.3	0.
40	0.5	0.5	0.
50	0.4	0.5	0.1
60	0.3	0.5	0.4
70	0.2	0.5	0.5
80	0.1	0.4	0.5
93	0.	0.2	0.3

9. v_g - settling velocity = 0.5 [m/day].

As leaf layers develop during simulation, LAI varied with leaf layers and with time (Table 9.5). The actual rate of deposition was proportional to LAI. If LAI became less than 25% of its maximum value after having passed that maximum, the deposition rate was 25% of δ.

The crop was inoculated at the centre of the field on the leaf layer closest to the soil at time T = 0 by 1 successful spore.

The field was represented by a grid of 11 x 11 x 7 points, or by 7 horizontal layers of 11 x 11 points. To exclude the influence of boundaries on the simulated results, the soil and the air above crop were represented by two horizontal layers each. Spores were dispersed in all directions and they were deposited on the five bottom layers. Spores deposited on the three middle layers, which represented the crop, could initialize lesions. As stated in Section 8.3.2., the appropriate boundary condition put the value of the spore density on the bottom layer equal zero. It simulated complete absorption of spores by a soil. The focus development was simulated during 87 days.

9.4.3 Results

Results were obtained by running the three-dimensional version of the 'diffusion model' on an Olivetti M280 personal computer, using PODESS (Appendix A). Disease severities at three times T = 71, 79, and 87 are shown in Fig. 9.11. They show a smooth but not a circular boundary of the focus. The centre of the focus has moved away from the centre of the field. The focus reaches different levels of disease severity on different leaf layers. The simulation results of Fig. 9.11 can be compared to the experimental field results of Fig. 9.12.

An interesting phenomenon can be observed, the delay in the shift of the focal centre relative to the time when the wind direction changed. This delay exists because the newly initialized lesions 'wait' for a latency period before they sporulate. For the same reason, waiting or delay periods due to latency, the spatial shift is smaller than might be expected. The delays in time and space of the shift of the focal centre are, together, a characteristic phenomenon which could be indicated as 'inertia' of the focus. The delay, though not unexpected, is demonstrated here for the first time by dynamic simulation.

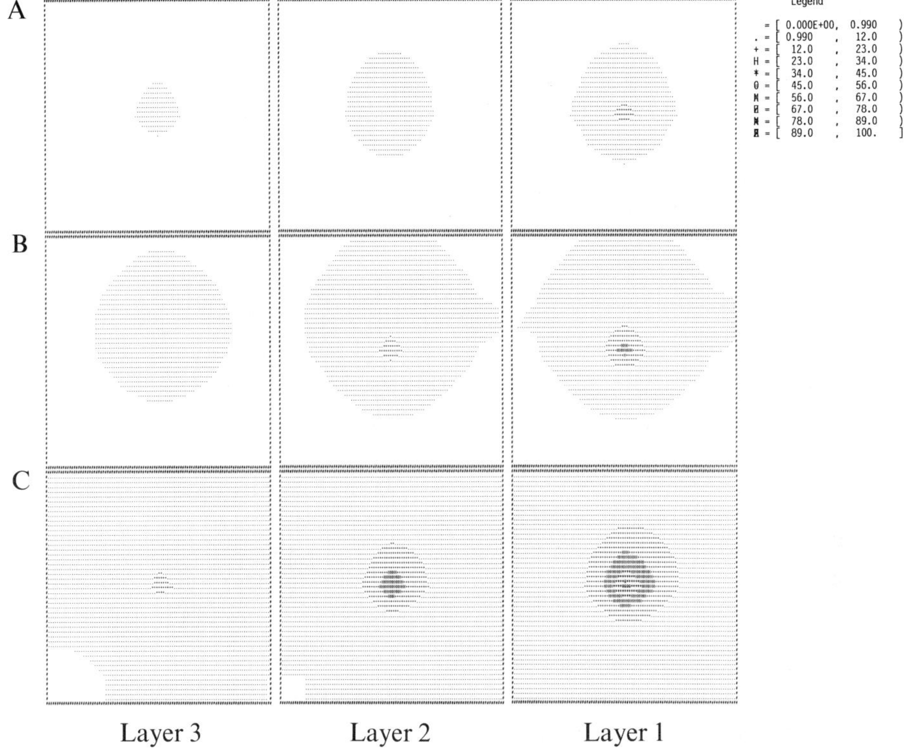

Fig 9.11. Focus expansion on leaf layers at three times: T = 71, 79, and 87. Results of the three-dimensional 'diffusion model' with variable leaf layers and variable wind. **A,** T = 71. **B,** T = 79. **C,** T = 87.

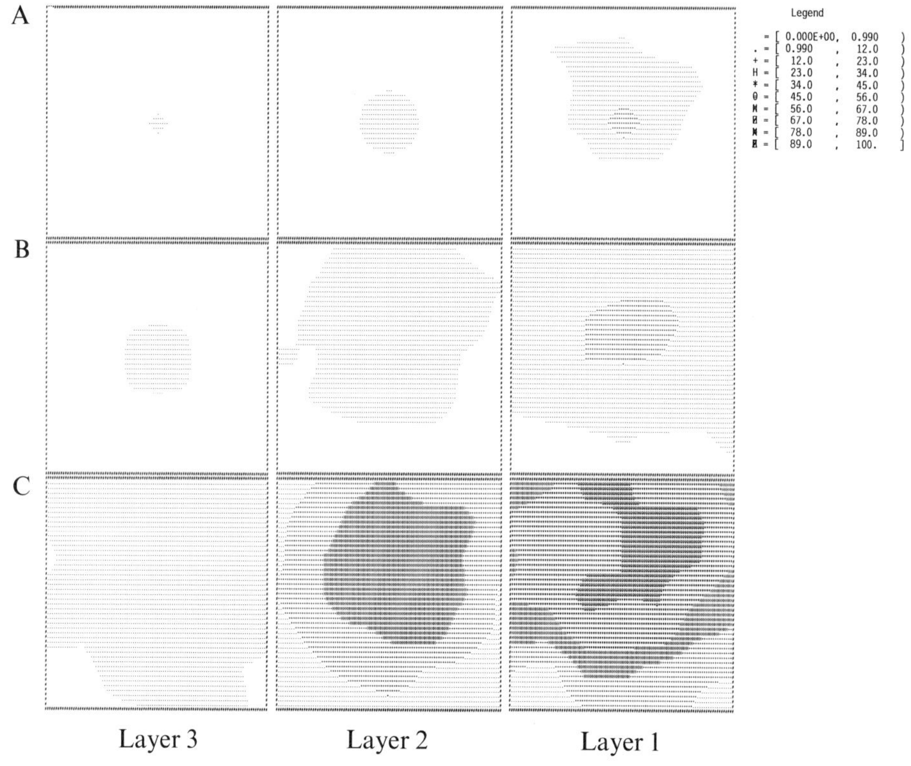

Fig 9.12. Focus expansion on leaf layers at three times: T = 71, 79, and 87. Experimental results of the leaf rust focus development on wheat (see Section 7.4). **A,** T = 71. **B,** T = 79. **C,** T = 87.

9.4.4 Discussion

The objective of running the three-dimensional `diffusion model' was the simulation of phenomena observed in the field: (1) different geometries of the focus on different leaf layers, and (2) a shift of the focal centre due to wind. Both real-life phenomena can indeed be simulated by the `diffusion model'. The quantitative differences between the simulated and the experimental results are due to the poor knowledge of parameter values and to the deterministic treatment inherent to the `diffusion theory'.

9.5 DISCUSSION

A mechanistic approach to focus formation leads to the `diffusion theory'. Expressed in mathematical terms as the system of equations (3.45), (3.46) it led to the `diffusion model', which is the basis for computer simulation. A great variety of simulation models can be constructed within the framework of the `diffusion theory' by appropriate modifications of parameters and/or of the original system of equations. The simulation models can be programmed in FORTRAN and handled by PODESS (Appendix A), which constitutes the numerical framework for applications of the `diffusion theory'.

Four models simulating focus formation under different conditions were discussed above. They are examples of possible applications to real-life epidemiological situations. The models allowed studying a few epidemiologically important phenomena, some already known, others newly discovered. New is the finding that the partitioning of spores over two dispersal mechanisms, one `short' and one `long', has an optimum for maximum disease development. This finding needs experimental verification.

Epidemiologists have, of course, often observed changes of foci under the influence of wind, though good descriptions and analyses are rare. New is the possibility of a quantitative, numerical description and explanation of field observations. The phenomena observed were: (1) a shift of the centre of the focus under the influence of wind and (2) a kind of `inertia' of the focus in reacting to the wind. Finally, well-known phenomena such as the `cryptic error' of field experiments (Van der Plank, 1963) and the appearance of daughter foci, can now be studied in detail. The simulation models based on the `diffusion theory' allow to estimate the numerical value of the `cryptic error', and to study the mechanism generating daughter foci.

10 GENERAL DISCUSSION

The 'diffusion theory' of focus development should be placed in an epidemiological context. The possibilities and deficiences of the approach to focus development in time and space, some ways of further improvement, and more fundamental extensions will be indicated.

10.1 THE PRESENT STATE OF EPIDEMIOLOGY

From the beginning, man struggled with the problem of feeding a growing population. The usual solution was to produce more, by methods such as increase of cropping area and of fertilizer dosage. Plant disease epidemics can spoil the gains achieved, because a higher crop area and a higher crop density lead to a higher number of sites available for infection. Then, the exponential phase of the 0 order epidemic (sensu Heesterbeek and Zadoks, 1987) will last longer. A higher density of fields leads to an easy spread of disease over a large area, resulting in a severe 1st order epidemic (sensu Heesterbeek and Zadoks, 1987). When a disease is given the opportunity to produce more inoculum, the starting level of the subsequent epidemic after a crop-free period may be high. Consequently, a severe 2nd order epidemic (sensu Heesterbeek and Zadoks, 1987) will develop.

As complete elimination of a disease from an ecosystem is virtually impossible, we must learn to live with it. A variety of methods is available, such as the application of chemicals, partial resistance of crops, eradication of inoculum sources, and so on, which help to keep disease severity below an economically harmful level (Zadoks, 1985; Zadoks and

Schein, 1979). These methods require much empirical knowledge, which takes years to collect. Extrapolation from existing knowledge to new situations can be facilitated by mathematics, which can summarize existing knowledge in relatively simple equations, allowing interpolation between the conditions for which the empirical experience exists. Results of equations can sometimes be extrapolated beyond the regions of experience, and thus lead to useful results. The greatest in this area was J.E. Vanderplank whose ideas and equations revolutionized plant pathology in the early sixties (Van der Plank, 1960; Zadoks and Schein, 1988). His ideas led to the first simple mathematical models applied to plant disease epidemics. Translated into the language of dynamic simulation (Zadoks, 1971; de Wit and Goudriaan, 1978; Rabbinge, 1982; de Wit, 1982), originally devised for engineers by Forrester (1961), these models gave many useful results about the development of disease in time. They led to warning systems like EPIPRE (Zadoks, 1988) which, predicting future levels of disease severity, allow to choose an economic way of plant protection.

New steps toward a general model of plant disease development were made in late seventies following two independent approaches. One approach was the extension of computer simulation models to cover disease development both in time and space (Kiyosawa, 1976; Kampmeijer and Zadoks, 1977). EPIMUL, developed by Kampmeijer and Zadoks (1977), was applied elsewhere in phytopathology (Mundt et al., 1986a, b, c). Combining Vanderplank's temporal development with a spatial dispersal mechanism, EPIMUL allows disease simulation in time and in non-uniform two-dimensional space. The second approach, derived from the integro-differential model of Kermack and McKendrick (1927) was developed independently by Diekmann (1978; 1979) and Thieme (1977; 1979). Starting from simple and general assumptions about disease, they produced an integro-differential equation, which describes spatio-temporal disease development in general terms. It can be adjusted to any particular situation by assuming specific dispersal and/or spore production mechanisms. The Diekmann-Thieme theory is the most general theory of plant disease development in time and space. However, being so general, the Diekmann-Thieme theory is difficult to apply in practice. Morever its main results are only valid asymptotically for large time, and under the restrictive conditions of a large field uniformly covered by a crop. The specialization of the Diekmann-Thieme theory needed to apply the results in a qualitative manner to phytopathological situations was provided by Van den Bosch et al. (1988a, b, c). These authors also discussed some of the quantitative predictions derived from their specialization of the general theory, as well as quantitative field test for two concrete host-pathogen systems.

10.2 THE 'DIFFUSION THEORY' OF FOCAL DISEASE DEVELOPMENT

The 'diffusion theory' tries to combine a theoretical and a simulation approach. Being a specialization of the Diekmann-Thieme theory, it has a thorough theoretical underpinning. Morever, the particular specialization chosen is backed up by concrete physical considerations. The 'diffusion theory' is mathematically formulated as a system of two partial differential equations, which can be solved numerically in any special situation providing useful information about that situation. Following EPIMUL, the 'diffusion theory' combines Vanderplank's temporal model of disease development with a spatial model of spore dispersal. As spatio-temporal disease development is formulated by means of equations instead of distribution functions, the `diffusion theory' is more flexible and it allows more changes and extensions than most other simulation models in plant pathology. It should be seen as a general framework, which can be used to build special model for particular applications.

Being less abstractly general than the Diekmann-Thieme theory, the `diffusion theory' can be applied more easily. The power of the `diffusion theory' lies not only in the flexibility of the theory but also in the flexibility of the accompanying software, PODESS. Applied to a real-life situation, the `diffusion theory' gives the `diffusion model' for this situation, which programmed in FORTRAN and linked to PODESS provides a simulation model specific for that situation.

The simulation models derived from the `diffusion theory' allow to simulate many situations of epidemiological interest. Focus development in a non-uniform crop, multiple spore dispersal, stochasticity of lesion initialization, and wind effects all have been discussed above. Their study by means of appropriate `diffusion models' led to the explanation or discovery of phenomena which were known empirically or were not known at all. The possibilities (1) to calculate the value of the `cryptic error' (Van der Plank, 1963), (2) to partition spores between the `long' and the `short' spore dispersal for a maximum number of lesions in a field, (3) to generate daughter foci, and (4) to shift the centre of the focus with wind, can serve as examples. The sensitivity analysis applied to the `diffusion model' and the simulation of the double dispersal mechanism led to the examination of the influence of a finite site density and a finite field size on the `behaviour' of disease in foci.

10.3 POSSIBLE IMPROVEMENTS

The price to be paid for the generality and the flexibility of the `diffusion theory' and PODESS is the computer time needed by the `diffusion model', applied to a situation of practical interest. Thus, the `diffusion theory' should be solved by PODESS only at the stage of building the `diffusion model', when new problems and opportunities are encountered. If the `diffusion model' is shown to be appropriate and applicable, it may become profitable to use faster but less general methods to solve numerically the system of equations of the `diffusion theory'. Therefore, PODESS should be extended by fast methods of numerical solution of special types of partial differential equations.

The sensitivity analysis of Chaper 5 was applied to the `diffusion theory' only once for relatively short simulation runs. Longer duration of runs might give additional information about the effect of exhaustion of the sites available for infection on the diffusion model's response. All simulation models discussed above should be subject to sensitivity analysis for a more accurate examination of their results. This is especially true for the `diffusion model' simulating the double spore dispersal mechanism, as the maximum effect of spore partitioning should yet be examined in more detail.

The `diffusion theory' has been applied to a situation with three dimensions in space. Now, the double dispersal mechanism should be included. As a three dimensional model necessitates spatial variation of the diffusion coefficient, special numerical methods must be applied in the transition layer between the regions representing the crop and the air above it. The equations of the `diffusion theory' must be solved separately within the crop and within the air above crop regions and the solutions must be adjusted to each other by appropriate boundary conditions.

10.4 FUTURE DEVELOPMENTS

The `diffusion theory' was derived and validated for relatively small fields, where spore dispersal is within the micro- or short mesoscale (Zadoks and Schein, 1979). This situation leads to formation of a single focus or at most a few foci (the 0 order epidemic sensu Heesterbeek and Zadoks, 1987). But plant disease epidemics spread over large areas (long meso- and macroscale sensu Zadoks and Schein, 1979), infecting many fields (1st order epidemic sensu Heesterbeek and Zadoks, 1987) with transport of spores over medium and long distances. The necessity of considering long-distance dispersal was indicated by Zadoks and Schein (1979) and Jeger (1985a). Methods to monitor (Nagarajan and Singh, 1974; Nagarajan et al., 1976; Westbrook, 1985) and study (Knox, 1974; Nagarajan and Singh, 1975; Nagarajan, 1977; Ermak, 1977; Pedgley, 1982; van Egmond and Kesseboom, 1983; Pedgley, 1985; Sparks et al., 1985) long-distance dispersal have been proposed by various authors. Combination of such methods (applied to long-distance dispersal) with the `diffusion theory' (applied to short-distance dispersal) seems to be in the range of possibilities.

The introduction or appearance of new pathogens or strains in areas where they did not exist before, may lead to the problem indicated as `crop vulnerability' (Horsfall at al., 1972). Zadoks and Kampmeijer (1977) raised the question whether crop vulnerability could be quantified and gave a tentative solution. `Diffusion models' could be used to the same purpose. They also permit to estimate effects of field size, distance between fields, gene development, and intercropping.

Plant resistance is often of short duration due to

mutations in the pathogen. Simulation of this process
might be possible by a joint solution of the equations
describing genetical changes in populations (Crow and
Kimura, 1970) and of the equations of the `diffusion
theory'. As PODESS is designed to solve an arbitrary
system of partial differential equations this joint
solution is relatively easy, at least with respect to
the available software.

The `diffusion theory' was designed to model the
focal development of foliar plant diseases caused by
air-borne fungi. The diffusion equation (3.46) was
derived with assumptions which hold only for these
diseases. Nevertheless, the dispersal of other
pathogens and pests can be also described
approximatively by the diffusion equation. An example
is the dispersal of beetles (Wetzler and Risch, 1984).
Thus, the framework offered by the `diffusion theory'
could be used to mimick dispersal of these agents,
extending the region of applicability of the `diffusion
theory'.

Following Van der Plank (1963), we repeat that
`epidemiological analysis has come to stay'. The
`diffusion theory' proposed in the present volume is
another contribution, we hope, to plant disease
epidemiology, which was given a position of prominence,
over a quarter of a century ago, by J.E. Vanderplank.

SUMMARY

Chapter 1. The 'diffusion theory' of focus development in plant disease is introduced. Foci develop in space and time. The theory applies primarily to air-borne fungal diseases of the foliage.

Chapter 2. The contents of the present volume are outlined.

Chapter 3. The 'diffusion theory' of focus development, intended to model phytopathologically interesting phenomena, emerges from a simple set of assumptions summarizing existing knowledge of plant pathologists. Using relatively easy and clear inferences, supported by methods used in mathematics and physics, this knowledge leads to a system of two partial differential equations (3.45), (3.46). These equations represent the 'diffusion theory' in mathematical terms.

Chapter 4. As any other new theory, the 'diffusion theory' must be validated by comparing it to models known from the literature and to experimental results. The 'diffusion theory' is a theoretical construct, permitting the development of a dynamic simulation model here indicated as the 'diffusion model'. The 'diffusion theory' is validated by comparing the 'diffusion model' to the model of Minogue and Fry and to EPIMUL of Kampmeijer and Zadoks. The various models show good qualitative and fair quantitative consistency. The quantitative differences are due to different assumptions about spore dispersal and deposition mechanisms. More important, the 'diffusion theory' was successfully validated by comparing its predictions to experimental data from yellow stripe rust (*Puccinia striiformis*) on wheat and from downy mildew (*Peronospora farinosa*) on spinach.

Chapter 5. Having derived the `diffusion model', it is necessary to determine its general behaviour for various parameter values. A method of sensitivity analysis, new to phytopathology, allows a detailed examination of linear, quadratic and mixed influences of parameters on responses of the `diffusion model'. The analysis indicated a few phytopathologically interesting relationships. As the `diffusion theory' attempts to describe the reality of plant disease dispersal in foci, the theory may lead to new hypotheses susceptible to experimental verification.

Chapter 6. The `diffusion theory' is based on an idealized picture of spore movement; spore motion is purely random at an infinitesimally small scale of time and space. Reality is different. Therefore, the `diffusion theory' is compared to the `telegrapher's theory', more complex and derived from different assumptions about spore motion. Comparison of the two theories does not indicate differences of practical importance, so that the diffusion approximation seems to be adequate for phytopathological applications.

Chapter 7. Wind is an important meteorological factor, affecting the development of air-borne plant disease. The extension of the `diffusion theory' to situations with a prevailing wind direction is made. The extended `diffusion theory' allows to build a `diffusion model' which simulates focus development under the influence of wind. The results of computer simulations show a good qualitative consistency with experimental data from brown leaf rust (*Puccinia recondita*) on wheat. In both cases, two phenomena were observed, a shift of the centre of the focus in the prevailing wind direction and, simultaneously, a certain `inertia' of the centre of the focus.

Chapter 8. The `diffusion theory', being formulated in terms of partial differential equations, can easily be extended by modification of parameters and/or equations. Thus, the appropriate `diffusion model' for any specific situation can be created on the basis of the `diffusion theory'. Application of the `diffusion theory' by means of a numerical solution of the apropriate `diffusion model' allows to study various characteristics of focal epidemics. Possibilities are discussed to combine the `diffusion theory' with some of the methods of computer simulation.

Chapter 9. Some results are given of models simulating phytopathologically interesting situations. These models allow to explain some real-life phenomena; the generation of daughter foci, the calculation of the `cryptic error' in plant breeding trials, and the interaction between two different mechanisms of spore dispersal.

Chapter 10. The book concludes with a general discussion. It shows the place of the `diffusion theory' among other models in plant pathology and it proposes some improvements and further developments of the theory.

Final remark. Mathematically, the `diffusion theory' is a special case of a more encompassing family of models considered by Diekmann and Thieme. The general results from the Diekmann-Thieme theory have been applied by Van den Bosch et al. in a phytopathological context using what amounts to a special limiting variant of the `diffusion theory'. These results only relate to the behaviour of the focal front for a large period of time in an area which is homogeneously planted in all directions. The strength of the `diffusion model', as implemented, is its ability to deal also with short periods and with environments which vary in time and/or space.

SAMENVATTING

Hoofdstuk 1. De "diffusie theorie" van de
haardvorming bij planteziekten wordt ingeleid in
Hoofdstuk 1. Haarden groeien in ruimte en tijd. De
theorie is vooral van toepassing op anemochore
schimmelziekten van het loof.

Hoofdstuk 2 geeft een overzicht van de inhoud van
dit boekwerk.

Hoofdstuk 3. De "diffusie theorie" van de
haardvorming in ruimte en tijd, ontworpen om
verschijnselen van fytopathologisch belang te
modelleren, wordt ontwikkeld uit een reeks van aannamen
die de bestaande fytopathologische kennis samenvatten.
Relatief eenvoudige en duidelijke redenaties,
ondersteund door methoden in gebruik bij de wiskunde en
de natuurkunde, leiden tot een stelsel van twee
partiele differentiaalvergelijkingen, (3.45) en (3.46).
Deze vergelijkingen geven de "diffusie theorie" weer in
wiskundige vorm.

Hoofdstuk 4. De "diffusie theorie" moet, net als
iedere andere nieuwe theorie, worden gevalideerd door
haar te vergelijken met modellen uit de literatuur en
met proefresultaten. De "diffusie theorie" leidt tot de
ontwikkeling van een dynamisch simulatiemodel hier
aangeduid als "diffusie model". De "diffusie theorie"
werd gevalideerd door vergelijking van het "diffusie
model" met het model van Minogue en Fry en met EPIMUL
van Kampmeijer en Zadoks. De verschillende modellen
tonen een goede kwalitatieve en een redelijke
kwantitatieve overeenkomst. De kwantitatieve
verschillen kunnen toegeschreven worden aan verschillen
in aannamen inzake verspreiding en depositie van
sporen. De "diffusie theorie" is met succes gevalideerd
door toetsing van haar voorspellingen aan

proefresultaten met gele roest (*Puccinia striiformis*)
op tarwe en valse meeldauw (*Peronospora farinosa*) op
spinazie.

Hoofdstuk 5. Als het "diffusie model" afgeleid is,
moet zijn algemeen gedrag bepaald worden in
afhankelijkheid van een aantal parameter-waarden. Een
gevoeligheidsanalyse, die nieuw is voor de
fytopathologie, maakt een gedetailleerd onderzoek
mogelijk van lineaire, kwadratische en gemengde
invloeden van parameters op de resultaten van het
"diffusie model". De analyse wees op enkele
fytopathologisch interessante verbanden. Aangezien de
"diffusie theorie" probeert de werkelijkheid te
beschrijven bij de verspreiding van planteziekten in
haarden kan deze theorie leiden tot nieuwe,
experimenteel verifieerbare hypothesen.

Hoofdstuk 6. De "diffusie theorie" gaat uit van een
geidealiseerd beeld van de beweging van sporen; deze is
volledig door het toeval bepaald op een oneindig kleine
schaal van tijd en ruimte. De werkelijkheid is anders.
Daarom wordt de "diffusie theorie" vergeleken met de
"telegrapher's theorie", die voor een speciaal, meer
gedetailleerd model, dat in de limiet tot een diffusie
vergelijking leidt, nog een extra correctieterm
meeneemt. Vergelijking van de beide theorieën wijst
niet op verschillen van praktisch belang, zodat de
diffusie benadering toereikend lijkt voor
fytopathologische toepassingen.

Hoofdstuk 7. Wind is een belangrijke meteorologische
factor, die de ontwikkeling van anemochore
planteziekten beinvloedt. De "diffusie theorie" wordt
uitgebreid tot situaties met een overheersende
windrichting. De uitgebreide "diffusie theorie" maakt
het mogelijk een "diffusie model" te bouwen dat
haardvorming onder de invloed van wind simuleert. De
resultaten van computer simulaties tonen een goede
kwalitatieve overeenstemming met proefresultaten van

bruine roest (*Puccinia recondita*) op tarwe. In beide gevallen werden twee verschijnselen waargenomen, een verschuiving van het centrum van de haard in de richting van de heersende wind en, tegelijkertijd, een zekere "traagheid" van het centrum van de haard.

Hoofdstuk 8. Aangezien de "diffusie theorie" geformuleerd is in termen van partiële differentiaalvergelijkingen, kan zij gemakkelijk uitgebreid worden door wijziging van parameters en/of vergelijkingen. Zo kan een passend "diffusie model" gemaakt worden voor iedere specifieke situatie, uitgaande van de "diffusie theorie". Toepassing van de "diffusie theorie" door middel van de numerieke oplossing van een geschikt "diffusie model" maakt het mogelijk diverse eigenschappen van focale epidemieën te bestuderen. Mogelijkheden worden besproken om de "diffusie theorie" te combineren met enkele methoden van computer simulatie.

Hoofdstuk 9. Enkele resultaten worden vermeld van modellen, die fytopathologisch interessante situaties simuleren. Deze modellen maken de verklaring mogelijk van enkele levensechte verschijnselen, zoals de verwekking van dochterhaarden, de berekening van de "verborgen fout" in proeven van planteveredelaars, en de interactie tussen twee verschillende mechanismen van sporenverspreiding.

Hoofdstuk 10. Het boek besluit met een algemene discussie. Deze bespreekt de plaats van de "diffusie theorie" temidden van andere modellen in de planteziektenkunde en doet voorstellen voor verbetering en voortgezette ontwikkeling van de theorie.

Slotopmerking. Wiskundig bezien is de "diffusie theorie" een bijzonder geval van een meer omvattende familie van modellen bestudeerd door Diekmann en Thieme. De analytische resultaten van de Diekmann-Thieme theorie zijn in een fytopathologische context toegepast door van den Bosch et al. met

gebruikmaking van een limiet variant van de "diffusie
theorie". De analytische resultaten van Diekmann en
Thieme en van van den Bosch hebben alleen betrekking op
het gedrag van het front van de haard over lange
tijdspannen in een in alle richtingen homogeen beplant
vlak. De kracht van het "diffusie model" is zijn
toepasbaarheid op korte tijdspannen en op milieu's die
varieren in tijd en/of ruimte.

REFERENCES

Ames, W.F., 1977. Numerical methods for partial differential equations. New York, Academic Press. 365 pp.

Anonymous, 1953. Some further definitions of terms used in plant pathology. Transactions of the British Mycological Society 36: 267.

Ash, M., 1979. Nuclear reactor kinetics. New York, McGraw-Hill Book Company. 450 pp.

Aylor D.E., 1978. Dispersal in time and space: Aerial pathogens, 159 - 180. In: J.G. Horsfall and E.B. Cowling (Eds). Plant Disease, an Advanced Treatise, Volume II, How Disease Develops in Populations. New York, Academic Press. 436 pp.

Aylor D.E., 1987. Deposition gradients of urediniospores of *Puccinia recondita* near a source. Phytopathology 77: 1442 - 1448.

Aylor D.E., H.A. McCartney and A. Bainbridge, 1981. Deposition of particles liberated in gusts of wind. Journal of Applied Meteorology 20: 1212 - 1221.

Buiel, A.A.M., M.A. Verhaar, F. van den Bosch, W. Hoogkamer and J.C. Zadoks, 1989. The effect of variety mixtures on the expansion velocity of yellow stripe rust foci in winter wheat, (Synopsis). Netherlands Journal of Agricultural Science.

Barrett J. A., 1980. Pathogen evolution in multilines and variety mixtures. Journal of Plant Diseases and Protection 87: 383 - 396.

Box G.E.P. and J.S. Hunter, 1957. Multifactor experiment designs for exploring response surfaces. Annals of Mathematical Statistics 28: 195 - 241.

Broadbent S.R. and D.G. Kendall, 1953. The random walk of *Trichostrongylus retortaeformis*. Biometrika 9: 460 - 465.

Cammack, R.H., 1958. Factors affecting infection gradients from a point source of *Puccinia polysora* in a plot of Zea mays. Annals of Applied Biology 46: 186 - 197.

Carver, M.B., D.G. Stewart, J.M. Blair and W.N. Selander, 1978. The FORSIM IV simulation package for the automated solution of arbitrary defined partial and /or ordinary differential equations systems. Chalk River Nuclear Laboratories - Atomic Energy of Canada Ltd. AECL - 5821. 154 pp.

Chamberlain, A.C., 1967. Deposition of particles to natural surfaces, 138 - 164. In: P.H. Gregory and J.L. Monteith (eds). Airborne Microbes. Seventeenth Symposium of the Society for General Microbiology held at The Imperial College, London, April 1967. London, Cambridge University Press. 385 pp.

Chamberlain, A.C. and R.C. Chadwick, 1972. Deposition of spores and other particles on vegetation and soil. Annals of Applied Biology 71: 141 - 158.

Cochran W. G., G. M. Cox, 1957. Experimental designs. Wiley, New York; Chapman and Hall, London. 611 pp.

Cox D. R., 1958. Planning of experiments. Wiley and Sons, New York; Chapman and Hall, London. 308 pp.

Crow J.F. and M. Kimura, 1970. An introduction to population genetics theory. Harper and Row, New York. 591 pp.

DEC, 1982. VAX/VMS DCL dictionary. Part 1: using the command language, September 1984. Marynard, Massachusetts, Digital Equipment Corporation. 121 pp.

DEC, 1984. Programming in VAX FORTRAN. September 1984. Marynard, Massachusetts, Digital Equipment Corporation.

De Wit C.T., 1982. Simulation of living systems, 3 - 8. In: Penning de Vries F.W.T. and H.H. van Laar (Eds.), Simulation of plant growth and crop protection. Wageningen, Pudoc. 308 pp.

De Wit C.T. and J. Goudriaan, 1978. Simulation of ecological processes. Wageningen, Pudoc. 175 pp.

Diekmann, O., 1978. Thresholds and travelling waves for the geographical spread of infection. Journal of Mathematical Biology 6: 109 - 130.

Diekmann, O., 1979. Run for your life. A note on the asymptotic speed of propagation on an epidemic. Journal of Differential Equations 33: 58 - 73.

Draper N. R., H. Smith, 1966. Applied Regression Analysis. New York, Wiley. 407 pp.

Eisensmith, S.P., R. Rabbinge and J.C. Zadoks, 1985. Development of a stochastic spore germination model. Netherlands Journal of Plant Pathology 91: 137 - 150.

Ermak, D.L., 1977. An analytical model for air pollutant transport and deposition from a point source. Atmospheric Environment 11: 231 - 237.

Finney D. J., 1960. An Introduction to the Theory of Experimental Design. University of Chicago Press, Chicago. 223 pp.

Forester, J.W., 1965. Industrial dynamics. Massachusetts Institute of Technology Press, Cambridge (USA). 464 pp.

Gear, C.W., 1971. Numerical initial value problems in ordinary differential equations. Prentice Hall, Englewood Cliffs, N. J. 253 pp.

Glasstone, S. and M.C. Edlund, 1956. The elements of nuclear reactor theory. Princeton, Van Nostrad Company. 416 pp.

Goudriaan, J., 1977. Crop micrometeorology: a simulation study. Wageningen, Pudoc. 249 pp.

Gregory, P.H., 1968. Interpreting plant disease dispersal gradients. Annual Review of Phytopathology 6: 189 - 212.

Gregory, P.H., 1973. The microbiology of the atmosphere, 2nd edition. Aylesbury (Bucks), Leonard Hill. 377 pp.

Hallam, T.G. and S.A. Levin (Eds), 1986. Mathematical ecology. Berlin, Springer. 457 pp.

Heesterbeek J.A.P. and J.C. Zadoks, 1987. Modeling pandemics of quarantine pests and diseases: problems and perspectives. Crop Protection 6: 211 - 221.

Hindmarsh, A.C., 1974. GEAR: Ordinary differential equation system solver. UCID - 30001 Rev. 3. Lawrence Livermore Lab., University of California, Livermore, Calif. (December 1974). 71 pp.

Horsfall, J.G. et al., 1972. Genetic vulnerability of major crops. National Academy of Science. Washington, D.C. 307 pp.

Ingold, C.T., 1967. Liberation mechanisms of fungi, 102 - 115. In: P.H. Gregory and J.L. Monteith (eds). Airborne microbes. Seventeenth Symposium of the Society for General Microbiology held at The Imperial College, London, April 1967. London, Cambridge University Press. 385 pp.

Jeger, M.J., 1985a. Long distance transport of aerially dispersed fungal pathogens, 107 - 113. In D.R. MacKenzie, C.S. Barfield, G.G. Kennedy, R.D. Berger, and D.J. Taranto (eds.). The movement and dispersal of agriculturally important biotic agents. Claitor's Publishing Division, Baton Rouge. 611 pp.

Jeger, M.J., 1985b. Models of focus expansion in disease epidemics, 279 - 288. In D.R. MacKenzie, C.S. Barfield, G.G. Kennedy, R.D. Berger and D.J. Taranto (eds.). The movement and dispersal of agriculturally important biotic agents. Claitor's Publishing Division, Baton Rouge. 611 pp.

Jennrich R. I., 1977. Stepwise Regression, pp. 58 - 75. In K. Enslein, A. Ralston and 'H. S. Wilf: Statistical Methods for Digital Computers. New York, Wiley. 454 pp.

Kampmeijer, P. and J.C. Zadoks, 1977. EPIMUL, a simulator of foci and epidemics in mixtures of resistant and susceptible plants, mosaics and multilines. Wageningen, Pudoc. 50 pp.

Kato, H., 1974. Epidemiology of rice blast disease. Review of Plant Protection Research 7: 1-20.

Kendall, D.G., 1948. A form of wave propagation associated with the equation of heat conduction. Proceedings of the Cambridge Philsophical Society 44: 591 - 593.

Kendall, D.G., 1965. Mathematical models of the spread of infection, 213 - 225. In: Mathematics and computer science in biology and medicine. Medical Research Council, H.M. Stationary Office.

Kermack, W.O. and A.G. McKendrick, 1927. A contribution to thr mathematical theory of epidemics. Proceedings of the Royal Society A115: 700 - 721.

Kingsolver, C.H., E.C. Peet and J.F. Underwood, 1984. Measurement of the epidemiological potential of wheat stem rust: St. Croix, U.S. Virgin Islands, 1954-57. Bulletin 854. The Pennsylvania State University, College of Agriculture. 21 pp.

Kittel, C., W. D. Knight, M. A. Ruderman, 1965. Mechanics. Berkeley physics course: vol. 1. New York, McGraw-Hill. 426 pp.

Kiyosawa S., 1976. A Comparison by simulation of disease dispersal in pure and mixed stands of susceptible and resistant plants. Japanese Journal of Breeding 26: 137 - 145.

Knox, J.B., 1974. Numerical modeling of the transport diffusion and deposition of pollutants for regions and extended scales. APCA Journal, 24: 660 - 664.

Korn, G.A. and J.V. Wait, 1978. Digital continuous-system simulation. Englewood Cliffs N.J., Prentice-Hall. 212 pp.

MacKenzie, D.R., 1976. Applications of two epidemiological models for the identification of slow rusting in wheat. Phytopathology 66: 55 - 59.

Manczak K., 1976. Technika planowania eksperymentu. Warszawa, Wydawnictwa Naukowo Techniczne. 277 pp.

(in Polish).

McCartney, H.A. and B.D.L. Fitt, 1985. Construction of Dispersal Models, 107 - 143. In: C.A. Gilligan (Ed.), Advances in Plant Pathology, volume 3, Mathematical Modelling of Crop Disease. London, Academic Press. 255 pp.

McLean R. A., V. L. Anderson, 1984. Applied Factorial and Fractional Designs. New York, Marcel Dekker. 373 pp.

Mehta, Y.R. and J.C. Zadoks, 1970. Uredospore production and sporulation period of *Puccinia recondita* f. sp. *tritici* on primary leaves of wheat. Netherlands Journal of Plant Pathology 76: 267 - 276.

Minogue, K.P. and W.E. Fry, 1983. Models for the spread of disease: Model description. Phytopathology 73: 1168 -1173.

Minogue, K.P. and W.E. Fry, 1983 b. Models for the spread of disease: Some experimental results. Phytopathology 73: 1173-1176.

Mollison D., 1977. Spatial contact models for ecological and epidemic spread. Journal of the Royal Statistical Society B 39: 283 - 326.

Morse, P.M. and H. Feschbach, 1953. Methods of theoretical physics. New York, McGraw-Hill Book Company. 1978 pp.

Mosteller F., J. W. Tukey, 1977. Data Analysis and Regression. Reading, Addison-Wesley. 588 pp.

Mundt C.C., K.J. Leonard, W.M. Thal and J.H. Fulton, 1986 a. Computerized simulation of crown rust epidemics in mixtures of immune and susceptible oat plants with different genotype unit areas and spatial distributions of initial disease. Phytopathology 76: 590 - 598.

Mundt C.C. and K.J. Leonard, 1986 b. Analysis of factors affecting disease increase and spread in mixtures of immune and susceptible plants in computer-simulated epidemics. Phytopathology 76: 832 - 840.

Mundt C.C. and K.J. Leonard, 1986 c. Effect of host genotype unit area on development of focal epidemics of bean rust and common maize rust in mixtures of resistant and susceptible plants. Phytopathology 76: 895 - 900.

Nagarajan S., 1977. Meteorology and forecasting of epidemics of diseases. Symposium on Basic Sciences and Agriculture, 136 - 143. Indian National Science Academy, New Delhi.

Nagarajan S., H. Singh, 1974. Satellite television cloud photography - a new method to study wheat rust dissemination. Indian Journal of Genetics 34A: 486 - 490.

Nagarajan S., H. Singh, 1975. The Indian stem rust rules - an epidemiological concept on the spread of wheat stem rust. Plant Disease Reporter 59: 133 - 136.

Nagarajan S., H. Singh, L.M. Joshi and E.E. Saari, 1976. Meteorological conditions associated with long-distance dissemination and deposition of *Puccinia graminis tritici* urediniospores in India. Phytopathology 66: 198 - 203.

Notteghem, J.L. and G.M. Andriatompo. 1977. Mesure au champ de la résistance horizontale du riz à *Pyricularia oryzae*. Agronomie Tropicale 22: 400 - 412.

Okubo, A., 1980. Diffusion and ecological problems: mathematical models. Berlin, Springer. 254 pp.

Pedgley, D.E., 1982. Windborne pests and diseases. Meteorology of airborne organisms. Ellis Horwood Ltd., Chichester. 250 pp.

Pedgley D.E. 1985. Concepts of atmospheric science as they relate to the movement of biotic agents, 175 - 178. In D.R. MacKenzie, C.S. Barfield, G.G. Kennedy, R.D. Berger and D.J. Taranto (eds.). The movement and dispersal of agriculturally important biotic agents. Claitor's Publishing Division, Baton Rouge. 611 pp.

Petersen R. G., 1985. Design and Analysis of Experiments. New York, Marcel Dekker. 429 pp.

Plant Pathology Committee, 1950. Definitions of some terms used in plant pathology. Transactions of the British Mycological Society 33: 154 - 160.

Rabbinge R., 1976. Biological control of fruit-tree red spider mite. Wageningen. Pudoc. 228 pp.

Rabbinge R., 1982. Pests, diseases and crop production, 251 - 265. In: F.W.T. Penning de Vries and H.H. van Laar (Eds.), Simulation of plant growth and crop protection. Wageningen, Pudoc. 308 pp.

Rijsdijk, F.H. and R. Rappoldt, 1980. A model of spore dispersal inside and above canopies, 407 - 410. In: Federal Environmental Agency. The 1st international conference on aerobiology. Munich, 13-15 August, 1978. Erich Schmidt, Berlin. 471 pp.

Ralston, A., 1965. A first course in numerical analysis. New York, Mc Graw - Hill.

Schroedter H., 1960. Dispersal by air and water - the flight and landing, 169 - 227. In: J.G. Horsfall and A.E. Dimond. Plant Pathology. An advanced treatise. Volume III. Academic Press, New York. 675 pp.

Shampine, L.F. and M.K. Gordon, 1975. Computer solution of ordinary differential equations. The initial value problem. W.H. Freeman and Company, San Francisco. 318 pp.

Sparks A.N., J.K. Westbrook W.W. Wolf, S.D. Pair and J.R. Raulston 1985. Atmospheric transport of biotic agents on a local scale, 203 - 217. In: D.R. MacKenzie, C.S. Barfield, G.G. Kennedy, R.D. Berger and D.J. Taranto (eds.). The movement and dispersal of agriculturally important biotic agents. Claitor's Publishing Division, Baton Rouge. 611 pp.

Thieme, H.R., 1977. A model for the spatial spread of

an epidemic. Journal of Mathematical Biology 4: 337 - 351.

Thieme, H.R., 1979. Asymptotic estimates of the solutions of nonlinear integral equations and asymptotic speeds for the spread of populations. Journal fur die Reine und Angewandte Mathematik 306: 94 - 121.

Tyldesley, J.B., 1967. Movement of Particles in Lower Atmosphere, 18 - 30. In: P.H. Gregory and J.L. Monteith (eds). Airborne Microbes. Seventeenth Symposium of the Society for General Microbiology held at The Imperial College, London, April 1967. London, Cambridge University Press. 385 pp.

Van den Bosch F., J.C. Zadoks and J.A.J. Metz, 1988 a. Focus expansion in plant disease. I. The constant rate of focus expansion. Phytopathology 78: 54 - 58.

Van den Bosch F., J.C. Zadoks and J.A.J. Metz, 1988 b. Focus expansion in plant disease. II. Realistic parameter-sparse models. Phytopathology 78: 59 - 64.

Van den Bosch F., H.D. Frinking, J.A.J. Metz and J.C. Zadoks, 1988c. Focus expansion in plant disease III. Two experimental examples. Phytopathology 78: 919 - 925.

Van den Bosch F., J.A.J. Metz and O. Diekmann, (in prep.). The velocity of spatial population expansion.

Van der Plank, J. E., 1960. Analysis of epidemics, 229 - 289. In: J.G. Horsfall and A.E. Dimond (eds.). Plant Pathology. An advanced treatise. Volume III. New York, Academic Press. 675 pp.

Van der Plank, J. E., 1963. Plant diseases: epidemics and control. New York, Academic Press. 349 pp.

Vanderplank, J.E., 1975. Principles of plant infection. New York, Academic Press. 216 pp.

Van Egmond, N.D. and H. Kesseboom, 1983. Mesoscale air pollution dispersion models - II. Lagrangian puff model and comparison with Eulerian grid model.

Atmospheric Environment, 17: 267 - 274.

Weinberg, A.M. and E.P. Wigner, 1959. The physical theory of neutron chain reactors. The University of Chicago Press. 800 pp.

Westbrook J.K. 1985. Measurements and techniques to document aerobiological transport systems: 219 - 229. In D.R. MacKenzie, C.S. Barfield, G.G. Kennedy, R.D. Berger and D.J. Taranto (eds.). The movement and dispersal of agriculturally important biotic agents. Claitor's Publishing Division, Baton Rouge. 611 pp.

Wetzler, R.E. and S.J. Risch, 1984. Experimental studies of beetle diffusion in simple and complex crop habitats. Journal of Animal Ecology 53: 1 - 19.

Williams E.J., 1961. The distribution of larvae of randomly moving insects. Australian Journal of Biological Sciences 14: 598 - 604.

Wyld. H.W., 1976. Mathematical methods for physics. Readings, Benjamin Cummings. 628 pp.

Zadoks, J.C., 1958. Het gele-roestonderzoek in 1958. Tienjarenplan voor graanonderzoek Wageningen 5: 109 - 119. (in Dutch).

Zadoks, J.C., 1961. Yellow rust on wheat. Studies in epidemiology and physiologic specialization. Netherlands Journal of Plant Pathology (Tijdschr. Pl. Ziekten) 67: 69 - 256.

Zadoks, J.C., 1971. Systems analysis and dynamics of epidemics. Phytopathology 61: 600 - 610.

Zadoks, J.C., 1985. On the conceptual basis of crop loss assessement: The threshold theory. Annual Review of Phytopathology 23: 455 - 473.

Zadoks, J.C., 1988. EPIPRE, a computer-based decision support system for pest and disease control in wheat: Its development and implementation in Europe. Plant Disease Epidemiology 2: (in press).

Zadoks, J.C., A.O. Klomp and S.D. van Hoogstraten, 1969. Smoke puffs as models for the study of spore dispersal in and above a cereal crop. Netherlands Journal of Plant Pathology 75: 229 - 232.

Zadoks, J.C. and P. Kampmeijer, 1977. The role of crop populations and their deployment, illustrated by means of a simulator EPIMUL 76. Annals of the New York Academy of Sciences 287: 164 - 190.

Zadoks, J.C. and R.D. Schein, 1979. Epidemiology and plant disease management. New York, Oxford University Press. 427 pp.

Zadoks, J.C. and R.D. Schein, 1988. James Edward Vanderplank: maverick and innovator. Annual Review of Phytopathology 26: 31 - 36.

Zadoks, J.C. and J.A.G. van Leur, 1983. Durable resistance and host-pathogen-environment interaction, 125 - 140. In: F. Lamberti, J.M. Waller and N.A. van der Graaff (eds.). Durable resistance in crops. NATO ASI Series, Series A: Life Sciences 55, Plenum Press, New York. 454 pp.

Appendix A
PODESS version 3.53

A.1 INTRODUCTION

PODESS (Partial and/or Ordinary Differential Equations Systems Solver) is a package for solving an arbitrary system of partial and/or ordinary differential equations. The solution is performed by the method of lines. Space is discretized and spatial derivations are calculated by utilization of Lagrange interpolation polynomials (Carver et al., 1978). The user can apply an interpolation formula based on an arbitrary number of points from three up to the number of grid points in every direction and special retarded and advanced two-point formulas for the calculation of the first derivative for hyperbolic equations (Carver et al., 1978). After the calculation of the spatial derivatives, the equations are established by the user-supplied subroutine UPDATE. At this moment, every partial differential equation becomes a system of ordinary differential equations. The number of ordinary differential equations is equal to the number of the grid points. This system is solved by one of five integration methods: Euler, Runge-Kutta-Ralston rank 4 (Ralston, 1965), Runge-Kutta-Fehlberg rank 4 (Korn and Wait, 1978), Adams (Gear, 1971; Hindmarsh, 1974; Shampine and Gordon, 1975), or, for stiff problems (Gear, 1971; Hindmarsh, 1974). The methods of Adams and Gear are introduced by connecting to PODESS the package GEAR written by A.C. Hindmarsh (december 1974 version).

PODESS 3.53 is written in FORTRAN 77; it can be run on the VAX computer (DEC, 1982; DEC, 1984). Version 3.53 is working in time and/or up to three spatial dimensions. The package contains also input, output and plotting routines. The user has to link two or, for two

special cases, three subroutines to the package:
INITL with initial conditions.
UPDATE with equations and calls, output, and plot
      subroutines.
PEDERV with Jacobian values (only with Adams or Gear
      methods with user-supplied Jacobian option).
For his own special case, the user can add more FORTRAN
subroutines.

In scientific applications, on machines with a
4-byte representation of real numbers, the DOUBLE
PRECISION version is normally used (Shampine and
Gordon, 1975). This version is obtained by removing the
word C_DB_PR preceding DOUBLE PRECISION declarations
from the files PODESS.FOR, PODESSLB.FOR. User-supplied
subroutines should work also with DOUBLE PRECISION
values.

A.2 COMMUNICATION WITH THE PACKAGE

Communication of user-supplied subroutines with the
package is performed through COMMON blocks:
COMMON /INTEGT/ F(MAXODE)
     F   - matrix containing values of function to be
            solved (in subroutine INITL, the   user
            should give its initial values),
COMMON /DERVT/ FT(MAXODE)
     FT - matrix containing values of the   first
            derivative with time (in grid points),
For the above COMMON blocks the dimension MAXODE is
the maximum number of ordinary differential equations.
For VAX version MAXODE = 10000 and for PC version
MAXODE = 1000. The actual number of ordinary
differential equations equals the number of   the
ordinary differential equations defined by the user
plus those arising from the decomposition of partial
differential   equations,   where   every   partial
differential equation is replaced by a number of

ordinary differential equations equal to the number of
grid points in space. In user-supplied subroutines
dimensions should be equal to the number of all grid
points times the number of partial differential
equations plus the number of ordinary differential
equations.

COMMON /SPACEX/ INTRPX, NPOINX, XL, XR, LX, LXX, NEQDX,
                DX, X(101)

    all variables refer to the X-direction

    INTRPX   - number of points for interpolation
                   formula (default = 3),
    NPOINX   - number of grid points (default = 11),
    XL   - left end (default = 0.),
    XR   - right end (default = 1.),
    LX   - logical value; if .TRUE. the first
                   derivative is calculated (default =
                   .TRUE.),
    LXX   - logical value; if .TRUE. the second
                   derivative is calculated (default =
                   .TRUE.),
    NEQDX   - logical value; if .TRUE. then grid
                   points are not equidistant (default =
                   .FALSE.),
    DX   - distance of grid points (if NEQDX =
                   .FALSE.),
    X   - matrix containing values of grid points;
                   dimension 101 is the maximum default
                   value, should be equal to NPOINX,

COMMON /SPACEY/ INTRPY, NPOINY, YL, YR, LY, LYY, NEQDY,
                DY, Y(51)

    variables as in /SPACEX/ but now in the Y-direction,
    dimension 51 is the default maximum value, should be
    equal to NPOINY,

COMMON /SPACEZ/ INTRPZ, NPOINZ, ZL, ZR, LZ, LZZ, NEQDZ,
                DZ, Z(11)

    variables as in /SPACEX/ but now in the Z-direction,
    dimension 11 is the default maximum value, should be

equal to NPOINZ,

COMMON /ADVRET/ IX, IY, IZ

    IX   - middle point for advanced-retarded two
                      point interpolation formula in the
                      X-direction; for grid point values
                      smaller than IX interpolation is
                      advanced, for values greater than IX
                      interpolation is retarded (default = 0 -
                      first derivative is calculated by
                      INTRPX-point formula),

    IY   - as IX but in the Y-direction,

    IZ   - as IX but in the Z-direction,

COMMON /TIME/ T, DT, TOUT, TEND, TBGN, METHOD, ERMAX,
              ERMIN, DTMIN

    T   - variable representing time,

    DT   - time step; can be changed by
                      variable-step integrating methods,
                      initial value can be established by user
                      (default = 0.01),

    TOUT   - communication interval; every TOUT
                      results can be printed (default = 1.),

    TEND   - time of ending the simulation (default =
                      100.),

    TBGN   - time of beginning the simulation
                      (default = 0.),

    METHOD   - variable which determines method of
                      integration:
                      1 = Euler variable-step,
                      -1 = Euler fixed-step,
                      2 = Runge-Kutta-Ralston   rank   4
                           variable-step,
                      -2 = Runge-Kutta-Ralston   rank   4
                           fixed-step,
                      3 = Runge-Kutta-Fehlberg   rank   4
                           variable-step,
                      -3 = Runge-Kutta-Fehlberg   rank   4
                           fixed-step,

10, 11, 12, 13 =   variable-order   and
      variable-step Adams method,
20, 21, 22, 23 =   variable-order   and
      variable-step Gear method.
The second digit, for choosing the Adams
or Gear method, allows to indicate one
of the following options of corrector
iteration:
0 = fractional (fixpoint) iteration,
1 = chord method with   user-supplied
      Jacobian from PEDERV (see below),
2 = chord method with Jacobian generated
      internally,
3 = chord   method   with   diagonal
      approximation to Jacobian.
(default = 3),

ERMAX   - value of maximum error for integration;
              if error is greater than ERMAX, the time
              step is   halved   for variable-step
              methods, for fixed-step methods a
              warning message is printed at the end of
              the simulation (default = 1.E-5),

ERMIN   - value of minimum error for integration;
              if error is smaller than ERMIN, the time
              step is   doubled   for   variable-step
              methods (default   =   1.E-7)   (this
              parameter is used only with METHOD = 1,
              -1, 2, -2, 3, -3),

DTMIN   - value of minimum time step; if DT is
              smaller   than   DTMIN   simulation   is
              terminated (default = 1.E-7),

Subroutine PEDERV has the form:

SUBROUTINE PEDERV(N,T,Y,PD,N0)

This   subroutine   should   supply   the   partial

derivatives of f(y,t) with respect to y (i.e., the
Jacobian matrix), evaluated at T = t and Y = y. It
must form a two-dimensional array PD, stored as an
N0 x N0 array, according to

$$PD(i,j) = \frac{\partial f_i}{\partial y_j}, \qquad\qquad 1 \le i,j \le N .$$

COMMON /CONTRL/ INOUT, NODE, NPDE, NSTART, NSTEP,
                NFE, NJE, IER

    INOUT   - variable indicating state of simulation
                   (cannot be changed by the user):
                   0 = the   simulation   is   in   its
                       communication interval; output is
                       impossible,
                   1 = end   of   communication   interval;
                       output is possible,
                   2 = end of simulation; output and final
                       calculations are possible,
    NODE   - number   of   ordinary   differential
                   equations (default = 0),
    NPDE   - number of partial differential equations
                   (default = 0),
    NSTART   - control variable
                   = 0 - Gear and Adams method does not
                         start at the begining of every
                         communication interval from the
                         begining,
                   > 0 - every   communication   interval,
                         Adams and Gear methods start from
                         the begining,
                   (default = 1),
    NSTEP   - number of time-steps,
    NFE   - number of UPDATE evaluations,
    NJE   - number of Jacobian evaluations,
    IER   - error indicator used by PODESS,

COMMON /BVMAT/ B(4,2,MAXBD)

    B - matrix of which the elements describe the boundary conditions; boundary conditions may be described by one of two equations:

    ($) B1 * FX(b) + B2 * F(b) = B3

    ($$) FT(b) = B2

    where b means the evaluation of a function in a boundary point.

    For element B(I,J,K):

        K - number of partial differential equation,

        J - indicates boundary point:

            1 = left point,

            2 = right point,

        I - indicates parameters:

            1 = B1 in ($)

            2 = B2 in ($) or ($$)

            3 = B3 in ($)

            4 = value of B(4,J,K) indicates type of condition:

                0. = condition ($); B1, B2, B3 can be functions of time, B1 and B2 can not be zero together,

               -1. = F(b) constant, then B1, B2, B3 are constants too,

               -2. = no boundary conditions,

               -3. = condition ($$)

  (default B1 = 0., B2 = 1., B3 = 0., B(4,J,K) = -1.)

  For the VAX version MAXBD = 3334 and for the PC version MAXBD = 334.

Boundary conditions are sent to the package through /BVMAT/ only if equations are solved in one spatial dimension. If the equations are solved in two or three spatial dimensions, the boundary conditions are sent by the parameter list of subroutine PDER2 or PDER3, respectively.

COMMON /CONS/ C(100)

    C - matrix of values of constants which are read-in

by the input subroutine,

COMMON /PARS/ P(200)

    P - matrix of values of parameters which are read-in by the input subroutine,

COMMON /FPLX/ NFX(20)

    NFX - matrix of names (10-character) which are read-in by the input subroutine (they can be used as names of spatial variables),

COMMON /FPLT/ NFT(30)

    NFT - matrix of names (10-characrer) which are read-in by the input subroutine (they can be used as names of time variables).

Elements of matrices C and P are read in format 3(A10,G16.8). Format A10 may be used for descriptions of input data on input file, but it is not required by PODESS. Elements of matrices NFX and NFT are read in format 8A10. Indicated dimensions are the default maximum values.

COMMON /LUN/ LI, LO, LPR

    LI   - logical unit number of input,

    LO   - logical unit number of output,

    LPR - control variable (LOGICAL):

        = .TRUE. - output line contains 132 characters,

        = .FALSE. - output line contains 80 characters.

A.3 THE IMPORTANT PACKAGE SUBROUTINES

SUBROUTINE PDER1 (F, FX, FXX)

    This subroutine calculates first and second derivatives with space for one-dimensional (in space) case.

SUBROUTINE PDER2 (F, FX, FXX, FY, FYY, BX, BY)

    This subroutine calculates first and second derivatives with respect to the X and Y directions for the two-dimensional (in space) case. The matrices BX and BY should contain the boundary conditions in the X and Y directions, respectively. The meaning of their elements is analogous to the elements of the matrix B from COMMON /BVMAT/, but instead of the number of partial differential equations the number of grid points in the perpendicular direction is used. Consequently BX should be declared as BX(4,2,NPOINY) and BY as BY(4,2,NPOINX) (where NPOINX is the number of grid points in X-direction and NPOINY is the number of grid points in Y-direction). Every partial differential equation must have its own boundary value matrices.

SUBROUTINE PDER3 (F, FX, FXX, FY, FYY, FZ, FZZ, BX, BY, BZ)

This subroutine calculates first and second derivatives with respect to the X, Y and Z directions for the three-dimensional (in space) case. The matrices BX, BY and BZ should contain the boundary conditions in the X, Y and Z directions, respectively. The meaning of their elements is analogous to the elements of the matrix B from COMMON /BVMAT/, but instead of the number of partial differential equations the number of grid points in the perpendicular directions is used. Consequently BX should be declared as BX(4,2,NYZ), BY as BY(4,2,NXZ) and BZ as BZ(4,2,NXY), where NYZ = NPOINY * NPOINZ, NXZ = NPOINX * NPOINZ and NXY = NPOINX * NPOINY (where NPOINX is the number of grid points in the X-direction, NPOINY is the number of grid points in the Y-direction and NPOINZ is the number of grid points in the Z-direction). Every partial differential equation must have its own boundary value matrices.

SUBROUTINE FTZER1 (FT)

This subroutine puts values of FT matrix elements on the boundary equal to zero if the boundary condition is ($) (see COMMON /BVMAT/).

SUBROUTINE FTZER2 (FT, BX, BY)

Analogous to FTZER1, but for the two-dimensional case. The matrices BX and BY must be declared as BX(4,2,NPOINY) and BY(4,2,NPOINX), respectively (where NPOINX is the number of grid points in the X-direction and NPOINY is the number of grid points in the Y-direction).

SUBROUTINE FTZER3 (FT, BX, BY, BZ)

Analogous to FTZER1, but for the three-dimensional case. The matrices BX, BY and BZ must be declared as BX(4,2,NYZ), BY(4,2,NXZ) and BZ(4,2,NXY), respectively (NYZ = NPOINY * NPOINZ, NXZ = NPOINX * NPOINZ, NXY = NPOINX * NPOINY) (where NPOINX is the number of grid points in the X-direction, NPOINY is the number of grid points in the Y-direction and NPOINZ is the number of grid points in the Z-direction).

SUBROUTINE OUTRS1 (F, NAMEF, IPR)

Output subroutine;

```
F          - one dimensional matrix,
NAMEF      - 10-character name of output matrix,
IPR        - control variable:
                 IPR = 0 - 11 elements of X are printed,
                 IPR > 0 - all elements of X are printed.
```

SUBROUTINE OUTRS2 (F, NAMEF, IPR)

Output subroutine;

```
F          - two-dimensional matrix,
NAMEF      - 10-character name of output matrix,
IPR        - control variable:
                 IPR = 0 - 11 rows, 11 elements per row,
                           of X are printed,
                 IPR > 0 - all elements of X are printed,
```

SUBROUTINE OUTXZ (F, NPOINX, NPOINY, NAMEF, IPR)

Output subroutine;

```
F          - two-dimensional matrix,
NPOINX     - number of rows of F,
```

NPOINY   - number of columns of F,
NAMEF   - 10-character name of output matrix,
IPR   - control variable:
                   IPR = 0 - 11 rows, 11 elements per row,
                         of X are printed,
                   IPR > 0 - all elements of X are printed,
SUBROUTINE FPLOT1 (M, F1, F2, F3, F4, F5, X, NDIM, FMIN,
                   FMAX, XMIN, XMAX, TITLE, NAMEX,
                   NAMEC, LSCLF, LSCLX, INTERP)
This subroutine plots the values stored in the
matrix F versus the values stored in the matrix X.
The plot is performed by the printer.
M   - number of plotted matrices,
F1   - one-dimensional matrix containing
                values of the plotted function,
F2,F3,F4,F5   - same as F1, but if M < 5, then the
                5-M last matrices must be dummy
                parameters,
X   - one-dimensional matrix containing
                the values of the grid points,
NDIM   - dimension of the matrices F1, F2,
                F3, F4, F5 and X,
FMIN,FMAX   - minimum and maximum values of the
                matrices F1, F2 F3, F4 and F5
                respectively, if FMIN = FMAX the
                subroutine chooses its own values,
XMIN,XMAX   - same, but for the matrix X,
TITLE   - 10-character title of plot,
NAMEX   - 10-character name of the matrix X,
NAMEC   - matrix of 10-character names of the
                matrices F1,F2, F3, F4 and   F5
                (dimension M)
LSCLF   - control variable:
                = 0 - the matrices F1, F2, F3, F4,
                      F5 are plotted on a linear
                      scale,
                = 1 - the matrices F1, F2, F3, F4,

                    F5 are plotted on a
                    logarithmic scale,
                 = 2 - the matrices F1, F2, F3, F4,
                    F5 are plotted on a logit
                    scale,
LSCLX   - same but for matrix X,
INTERP   - control variable for interpolation:
                 = 0 - no interpolation between data
                    points,
                 = 1 - linear interpolation between
                    data points,
                 > 1 - interpolation between data
                    points by cubic splines.
SUBROUTINE FPLOT2 (F, X, NPOINX, Y, NPOINY, FMN, FMX,
                 BLV, NAMEF, NAMEX, NAMEY, LSCLF,
                 INTX, INTY)
This subroutine plots the values stored in the
matrix F versus the values stored in the matrixes X
and Y. Ten different intensities are used. The plot
is performed only by the printer.
F   - two-dimensional matrix,
X   - one-dimensional matrix containing values
                   of the grid points in the X-direction,
NPOINX   - number of elements of X,
Y   - the one-dimensional matrix containing
                   values of the grid points in the
                   Y-direction,
NPOINY   - number of elements of Y,
FMN,FMX   - minimum and maximum values of the matrix
                   F respectively; if FMN = FMX the
                   subroutine establishes its own values,
BLV   - level under which the values are
                   represented as blank fields,
NAMEF   - 10-character name of the matrix F,
NAMEX   - 10-character name of the matrix X,
NAMEY   - 10-character name of the matrix Y,
LSCLF   - control variable:

                    = 0 - F plotted on a linear scale,
                    = 1 - F plotted on a logarithmic scale,
                    = 2 - F plotted on a logit scale,
    INTX   - control variable:
                    = 0 - no interpolation between the data
                          points in the X-direction,
                    = 1 - linear interpolation between the
                          data points in the X-direction,
    INTY   - control variable:
                    = 0 - no interpolation between the data
                          points in the Y-direction,
                    = 1 - linear interpolation between the
                          data points in the Y-direction,

SUBROUTINE FPLOTT (M, F1, F2, F3, F4, F5, FMIN, FMAX,
                   TITLE, NAMEF, LSCLF)

Subroutine stores values of up to five variables
during the program run and plots them versus time at
the end of the run. Values are stored for every
communication interval. The subroutine can be called
many times, but the number of stored values can not
be greater than 2010. The names of the plotted
functions are sent by COMMON /FPLT/.

M   - number of plotted functions   of
                      time,
F1,F2,F3,F4,F5 - functions to plot,
FMIN,FMAX   - minimum and maximum values of the
                      plotted functions; if FMIN = FMAX
                      the subroutine establishes its own
                      values,
TITLE   - 10-character name of plot,
NAMEF   - matrix of 10-character names of the
                      functions F1, F2, F3, F4, F5,
LSCLF   - control variable:
                        = 0 - functions are plotted on a
                              linear scale,
                        = 1 - functions are plotted on a
                              logarithmic scale,

      = 2 - functions are plotted on a
      logit scale.

As the Adams and Gear methods are introduced by connecting the package GEAR to PODESS, the user can also apply its subroutines, and then utilize all their possibilities (detailed description is in Hindmarsh, 1974).

FUNCTION FLININ (X, XW, FW, N)

  Function calculates a value by linear interpolation.

X - a value of an independent variable for which the result is calculated,

XW - values of an independent variable in the grid points (must be stored in ascending or descending order),

FW - values of a dependent variable in the grid points,

N - number of grid points.

A.4 USER–SUPPLIED SUBROUTINES

Generally these subroutines should have the form (in the 2- dimensional case):

```
    SUBROUTINE INITL
    COMMON    /INTEGT/ F(...)
   1          /CONTRL/ INOUT, NODE, NPDE
   2          /CONS/...
   3          /PARS/...
   4          /SPACEX/ INTRPX, NPOINX
   5          /SPACEY/ INTRPY, NPOINY
   6          /INIUPD/ NXY
    NPDE = ...
    NODE = ...
    NXY = NPOINX * NPOINY
      .

      .

    DO 10 J = 1, NPOINY
```

```fortran
      DO 10 I = 1, NPOINX
10       F(I,J) = initial values
      .
      .

      RETURN
      END


      SUBROUTINE UPDATE
      CHARACTER*10 NAMEF, NAMEX, NAMEY, NAMET, TITLE
      DIMENSION FX(...), FXX(...), FY(...), FYY(...)
      COMMON   /INTEGT/ F(...)
     1         /DERVT/ FT(...)
     6         /CONTRL/ INOUT
     7         /SPACEX/ INTRPX, NPOINX, ...
     8         /SPACEY/ INTRPY, NPOINY, ...
     9         /PARS/ ...
     1         /CONS/ ...
     2         /INIUPD/ N
     3         /FPLT/ NAMET(4)
      .

      .
      DATA FMN, FMX, BLV /2*0., 1. /, LF, LLF/1, 0/
      DATA NAMEF, NAMEX, NAMEY /'function F',
     1    'coordin. X', 'coordin Y'/,
     2    TITLE /'Lesion den'/
      .

      .
      CALL PDER2 (F, FX, FXX, FY, FYY, N, BX, NPOINY,
     1    BY, NPOINX)
      DO 10 J = 1, NPOINY
        DO 10 I = 1, NPOINX
10       FT(I,J) = equations
      IF (INOUT .EQ. 0) RETURN
      CALL OUTRS2 (F, NAMEF, 0)
      CALL FPLOT2 (F, X, NPOINX, Y, NPOINY, FMN, FMX,
     1    BLV, NAMEF, NAMEX, NAMEY, 0, 1, 0)
      CALL FPLOTT (4, F(1, 1), F(2, 1), F(3, 1),
```

```
    1    F(1, 2), F5, FMN, FMX, TITLE, NAMET, 0)
     RETURN
     END
```

If METHOD = 11 or 21, the user must program the subroutine PEDERV:

```
     SUBROUTINE PEDERV (N, T, Y, PD, N0)
     DIMENSION PD(N0,N0)
     PD(1,1) = ...
     PD(1,2) = ...
       .
       .
     PD(N0,N0) = ...
     RETURN
     END
```

A.5 RUNNING PODESS 3.53

At present, the package PODESS 3.53 exists in two versions:
1. VAX version, which is working on the VAX computer of the Agricultural University in Wageningen,
2. PC version (compiled by RM FORTRAN ver 2.11), which can work on IBM PC/XT/AT or a compatible personal computer under the DOS operating system version 2.11 or later.

A.5.1 The VAX version

To run his program the user should use the following VAX commands (files are called DISEASE.ext; where .ext is .FOR, .OBJ or .EXE; $ is a prompt of VAX).
1. Editing:
$ EDIT DISEASE.FOR - editing the file containing the subroutines INITL, UPDATE, and PEDERV (if necessary).
2. Compilation:

$ FORTRAN DISEASE   - compilation of the FORTRAN-file
                        DISEASE.FOR (extension .FOR is
                        default); the compiled file has
                        default name DISEASE.OBJ.

3. Linking:

$ LINK/EXECUTABLE=DISEASE.EXE PODESS,DISEASE,PODESSLB
                /LIBRARY
                     - the file DISEASE.OBJ is linked to
                       the file PODESS.OBJ, the file
                       PODESSLB.OLB is searched to
                       resolve all the external
                       references (PODESSLB.OLB is the
                       compiled library of subroutines
                       used by PODESS 3.53). The output
                       file has default name DISEASE.EXE.

4. Running the program:

$ RUN DISEASE   - run the program stored on
                       DISEASE.EXE file.

The response of the program is:

Welcome in PODESS version 3.53
Input,Output:

which requires indication of input and output files.

Input can be:
   CON   = terminal (default),
   Filename   = input file on disk (default extension
                  of filename is .DAT).
Output can be:
   CON   = terminal (default),
   LPT   = output is stored on file filespec.LPT
                  (where filespec is the name, without
                  extension, of the input file; if the
                  input is from a terminal, then
                  filespec = PODESS) in a form suitable
                  for the printer (132 characters per

line, subroutine FPLOT2 overprints
lines for plotting different
intensities of points),

RES = output is stored in the file
filespec.RES (where filespec is the
name, without extension, of the input
file; if the input is from a
terminal, then filespec = PODESS) in
the form suitable for the
presentation on a terminal (80
characters per line, subroutine
FPLOT2 does not work),

Filespec.ext = output is stored in the file
filespec.ext according to the
following rules: ext = LPT- form of
output the same as for LPT, ext
different from LPT - form of output
the same as for RES, (default ext =
LPT). Filespec must be different from
CON, LPT, and RES.

A.5.2 The PC version

To run his program the user should perform the
following steps (files are called DISEASE.ext; where
.ext is .FOR, .OBJ or .EXE).

1. Editing:

edit DISEASE.FOR - editing the file containing
subroutines INITL, UPDATE, and
PEDERV (if necessary), where
`edit' stands for a name of an
arbitrary editor which produces an
ASCII file.

2. Compilation:

RMFORT DISEASE/I - compilation of the FORTRAN-file
DISEASE.FOR (extension .FOR is
default); the compiled file has

the default name DISEASE.OBJ (the
compiler switch /I results in
INTEGER*2 interpretation of
INTEGER variables; also the switch
/Y must be used on IBM AT or
compatible machines, it forces the
RMFORT compiler to produce INTEL
80286 code).

3. Linking (one from two forms):
PLINK86 FI PODESS,DISEASE OUT DISEASE
        LIB PODESSLB,SCREEN,RMFORT
                        - the file DISEASE.OBJ is linked to
                        the file PODESS.OBJ, the files
                        PODESSLB.LIB, SCREEN.LIB and
                        RMFORT.LIB are searched to resolve
                        all the external references
                        (PODESSLB.LIB is the compiled
                        library of subroutines used by
                        PODESS 3.53, SCREEN.LIB is the
                        compiled library of PC screen
                        management routines and RMFORT.LIB
                        is the RM FORTRAN standard
                        library). The output file has the
                        name DISEASE.EXE.
The compilation and linking instructions, presented
above, assume that the compiler (RMFORT), the linker
(PLINK86), the main program (PODESS) and the libraries
(PODESSLB, SCREEN, RMFORT) are on the default disk
drive and the default directory. If this is not the
case, these file-names should be preceeded by
appropriate disk drive or/and directory specification.
4. Running the program:
DISEASE   - run the program stored in the
                        DISEASE.EXE file.
As the response, the program shows its name and after
pressing an arbitrary key, displayes a window, where it

asks for the `working directory', the name of the
`input file' and the name of the `output file'.
The input can be:

    CON   = terminal (default),
    Filename   = input   file   on   disk   (default
                       extension of filename is .DAT).

The output can be:

    CON   = terminal (default),
    PRN or PRN: = printer,
    LPT1 or LPT1:   = printer number 1,
    LPT2 or LPT2:   = printer number 2 (if existing),
    LPT3 or LPT3:   = printer number 3 (if existing),
    LPT   = the output is stored in the file
                       filespec.LPT (where filespec is the
                       name, without extension, of the
                       input file; if the input is from a
                       terminal, then filespec = PODESS)
                       in a form suitable for the printer
                       (132   characters   per   line,
                       subroutine FPLOT2 overprints lines
                       for plotting different intensities
                       of points),
    RES   = the output is stored in the file
                       filespec.RES (where filespec is the
                       name, without extension, of the
                       input file; if the input is from a
                       terminal, then filespec = PODESS)
                       in the   form   suitable   to   be
                       presented on the terminal   (80
                       characters per line, subroutine
                       FPLOT2 does not work),
    Filespec.ext   = the output is stored in the file
                       filespec.ext according to   the
                       following rules: ext = LPT- form of
                       the output the same as for LPT, ext
                       different from LPT - form of the
                       output the same   as   for   RES,

(default ext = LPT). The filespec
must be different from CON, LPT,
and RES.

A.6 ORGANIZATION OF THE INPUT FILE

The title of the simulation run (80-character)
should be stored in the first line. Every next block of
the input file must be preceded by the name of the
COMMON block to which is referred. Every block must be
finished by 10 spaces in the name field (format A10).
The blocks are:
/CONS/

A list of constant values in format 3(A10,G16.8).
Format A10 is provided for the description of every
value. Information from this field is not used by
PODESS, except in the particular case where the
field contains 10 spaces, which means end of block.
Numbers should appear in the same order as in
COMMON/CONS/ in the subroutines INITL and/or UPDATE.

/PARS/

A list of parameter values. The description is
analogous to the /CONS/ description, but now the
values are sent to COMMON/PARS/.

/XGRID/

A list of grid points in the X-direction. The
description is analogous to the /CONS/ description,
but now the values are sent to the matrix X in
COMMON/SPACEX/. The number of values must be equal
to NPOINX.

/YGRID/

A list of grid points in the Y-direction. The
description is analogous to the /CONS/ description,
but now the values are sent to the matrix Y in
COMMON/SPACEY/. The number of values must be equal
to NPOINY.

/ZGRID/

A list of grid points in the Z-direction. The description is analogous to the /CONS/ description, but now the values are sent to the matrix Z in COMMON/SPACEZ/. The number of values must be equal to NPOINZ.

/FPLX/

A list of 10-character names in format 8A10 which are sent to COMMON/FPLX/.

/FPLT/

A list of 10-character names in format 8A10 which are sent to COMMON/FPLT/.

The data for a single run must be ended by the command:

/RUN/

which starts the run. The following lines can contain data in the form described above, stored for later runs. The end sequence of the input file must be a blank line and the command:

/STOP/.

The commands and the names of the blocks must start from column 1. The user can use his own input instructions in the supplied subroutines, but the title line and the command:

/RUN/

must precede his data on the input file and this file must be ended as stated above.

A.7 SENDING FILES FROM FLOPPY DISK TO VAX

If files are only on floppy disk, the user should send by KERMIT the following files to VAX: PODESS.FOR and PODESSLB.FOR. After sending, all files must be compiled and the library PODESSLB must be created. This can be done by using the following VAX/VMS commands, which are given in detail in DEC, 1982; DEC, 1984.

Compilation:

$ FORTRAN PODESS,PODESSLB

compiled files will be PODESS.OBJ, PODESSLB.OBJ. File
PODESSLB.OBJ should be transformed to the library
PODESSLB.OLB, by:
$ LIBRARY/CREATE PODESSLB PODESSLB

Appendix B
Absorption and scattering

The theory introduced here, including the derivation of the diffusion equation, uses some concepts which may not be familiar to phytopathologists. The present appendix introduces these concepts in a simple way.

The macroscopic cross section can be described in the following way. A marksman shoots at targets having a density g (g is a volume density of targets $[NL^{-3}]$). The probability of success in hitting a single target is proportional to the surface T of the target. The probability of success in hitting a target at all is, moreover, proportional to the target density. The product of $g \cdot T$ is called the 'macroscopic cross section' of the targets. The foregoing illustration referred to 3-dimensional space, the following paragraphs refer to a field crop seen as a 2-dimensional space.

The macroscopic cross section for absorption can be described more precisely in two-dimensional space. Imagine a flow of spores through an absorbing medium (Fig. B.1). Concentrate for the time being on the spore flux $\Phi(r,\theta)$ with $\theta = 0$. Inside the medium, the spore flux Φ in the x direction (the number of spores flowing in the x direction per unit length per unit time) decreases over the distance dx by:

$$d\Phi = - g \, l \, dx \, \Phi \tag{B.1}$$

where g is the area density of absorbing sites and l is the length of such a site (the 2-dimensional case is considered here).

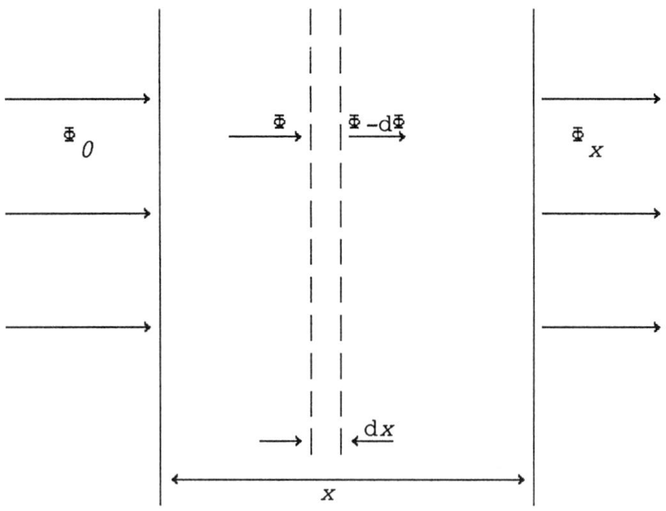

Fig. B.1. Spores fly through an absorbing medium. Spores from the incoming flux Φ_0 are absorbed during their travel along x. The outcoming flux is $\Phi_{x'}$ with $\Phi_x \leq \Phi_0$.

Dividing both sides of (3.2) by Φ and integrating from 0 to x we obtain:

$$\left[\, ln \, \Phi \, \right]_{\Phi_0}^{\Phi_x} = - g \, l \, x \qquad (B.2)$$

This is equivalent to:

$$\Phi_x = \Phi_0 \, \exp(-g \, l \, x) = \Phi_0 \, \exp(-C_a \, x) \qquad (B.3)$$

where $C_a = g \, l$ is the macroscopic cross section for absorption. C_a characterizes the absorbing medium. In a similar way the macroscopic cross section for other processes can be determined.

One of the other processes to be considered is scattering of spores (changing the direction of their movement) without changing their speed and without absorption. Scattering can be a result of air turbulence, wind gusts, or collisions with plant surfaces. The macroscopic cross section for scattering will be denoted as C_s.

CURRICULUM VITAE

Marek Zawołek was born October 21, 1955, in Cracow, Poland. In 1970, he entered secondary school, with an extended programme in mathematics and physics. In 1974, he entered the Jagiellonian University in Cracow, to study physics. In 1979, he received his MSc degree in physics. In April 1979, he started his work in the Cereals Department of the Institute for Plant Breeding and Acclimatization in Cracow, as a computer programmer, and from 1985, as the Head of the Biometrics Laboratory of the Cereals Department. From 1986, he also chairs the research unit `Application of statistics, mathematics and computer science to cereals breeding'. In 1983, the author participated in the Post-graduate Course "Simulation and Systems Management in Crop Protection" offered by the Wageningen Agricultural University. Various leaves of study in the Netherlands, financially supported by fellowships of FAO, the Dutch Ministry of Agriculture and Fisheries (via IAC), and the Wageningen Agricultural University, made the completion of this thesis possible.

ERRATA.

Page	line from top	is	should be
52	16, 17	infection efficiency	effectiveness
55	31	infection efficiency	effectiveness
57	29	infection efficiency	effectiveness
69	14	infection efficiency	effectiveness
80	4, 5	infection efficiency	effectiveness
81	6	infection efficiency	effectiveness
93	3	six	seven
93	10	infection efficiency	effectiveness